AGROECOLOGY

The Universal Equations

AGROECOLOGY

The Universal Equations

Paul A. Wojtkowski

CRC Press is an imprint of the
Taylor & Francis Group, an **informa** business
A SCIENCE PUBLISHERS BOOK

CRC Press
Taylor & Francis Group
6000 Broken Sound Parkway NW, Suite 300
Boca Raton, FL 33487-2742

First issued in paperback 2020

ISBN-13: 978-1-4987-4502-4 (hbk)
ISBN-13: 978-0-367-73744-3 (pbk)

Visit the Taylor & Francis Web site at
http://www.taylorandfrancis.com

and the CRC Press Web site at
http://www.crcpress.com

'Now pees' quod Nature 'I comaunde here'
('Peace now, I command here' said Nature)
Chaucer (1382) Parliament of Foules

Preface

It is commonly accepted that agriculture has existed for over 10,000 years. For most of this period, humans learned how to produce food with and without natures help. Agroecology was major factor, but went unrecognized.

This continued until the mid-1800s when science and the scientific method came to the fore. This was a period when, through numerous technological advances, there was a belief that mankind could control natural forces and eventually dominate nature.

Agriculture was at a disadvantage. Ecology was in its early infancy, lacking many of the key concepts. Statistics was a new science with little grounding. Given these limitations and with an inability to make headway against bio-complexity, early researchers sought simplification.

What resulted was a focus on the monoculture. Problems were best handled one-on-one. This favored chemicals. In time, genetics entered picture and, combined with a chemical reliance, this resulted what might be called the green revolution model.

The Green Revolution

The picture represented by the green revolution model is more than just high-yielding crops preforming at their maximum. This is a system where high yields are supported by large inputs of chemicals. Some are mildly harmful, others more worrisome.

These are also narrowly focused systems. Risk, inputs, costs, and environmental concerns are sidelined in the drive for yield and revenue-based profit maximization. For consumers, the enticement is food and fiber, readily available and cheaply purchased. The logical extension of extreme simplification is economies of scale and the factory farm. Unfortunately, in this drive for simplification, biodiversity and ecology are regarded as obstacles, not as means.

In this very constrictive, and somewhat basis, interpretation of the agricultural history, any deviation off this narrow track requires looking at biodiversity. Once this step is taken, it becomes apparent that much is missing. Roughly estimating, 90–95% of the agro-options and agro-possibilities are mostly ignored within the narrow confines of the green revolution model.

Alternative Agriculture

The agricultural mainstream, in continuing with the green revolution model, is falling short on any number of concerns. There are many that seek other options, other directions, and other methods. These are best if in-tune with nature and the environment. The rise of organic agriculture, permiculture, sustainable agriculture, and other departures from the agricultural norm are, in essence, cries for help. These demonstrates need to coalesce around some core alternative.

The Rise of Agroecology

Given what is happening and a growing chorus for something different, it is inevitable that agroecology will dominate. Agroecology offers alternatives to monocropping, agrochemicals, and the green revolution model. Moreover, from a scientific perspective, comparison of the rich theoretical core of ecology and agroecology to the narrow conceptional base of the green revolution model shows that much is lacking.

Whether or not one is in agreement with the latter statement, there are suppositions that can be drawn. As the agroecology expands, agriculture should become a subset of agroecology. The more likely course is that, as an academic discipline, agroecology will slowly supersede or absorb agriculture.

Because agroecology is able to go beyond the one-chemical, one-answer mindset, it is possible to embrace bio-complex, nature-friendly solutions. This is also because agroecology, in league with ecology, has the theoretical base to handle the various forms of biocomplexity.

In going into the bio-solutions, the emphasis in this text is not on expanding the field-level operational base, i.e., presenting and describing more of the agroecological options and advances. The goal is to intensify the theory that does exist and extend this to the furthest reaches.

Too Much Too Soon

For many, agroecology is a foreign creature. Ecologists, in stepping across into agroecology, find that their methods and results are now dictated by economics. For those in conventional agriculture, forestry, and agroforestry, the notion of being theory driven, rather that empirically derived, can be a radical departure from the previous norm. Again, for those that dispute the notion that agriculture is empirically driven, one only has to step beyond the simplicity of the green revolution model.

There are other foreign elements. Agroecology is relatively new. Instead of menu with few items, agroecology represents a large smorgasbord, tables laden with concepts, ideas, principles, practices, and approaches. Until such time as these are judged and thinned, practitioners must face an overly complex array of basic underpinnings and cropping alternatives.

For the overwhelmed practitioner, science and theory are not the only course. There are evolving guidelines that help formulate or manage classes of agroecosystems. Throughout this text, these are presented as an intermediate step between practice and theory.

Presented in this text is a vision on how those parts that constitute agroecology fit together. Some will accept, other may reject this continuum.

Hopefully, other ideas will follow and a lively debate will ensue (in as far as this is possible in agroecology). It is better if agroecology does not become defined by a belief or philosophy. Instead, agroecology, at its apex, should quantitatively express what nature contributes.

Uncommon Agroecosystems

Mentioned throughout this text are various seemingly alien agrosystems. For those vested in mono-cropping, many may seem beyond the pale. The many photos ground the book. Instead of pure abstraction and pure theory, the reality is that the agroecosystems mentioned exist. Admittedly, some are uncommon, of specialized use, and/or are only found in far regions.

Vocabulary

In a world that has grown accustomed to the terminology of conventional agriculture, there are difficulties in modifying the language to address an expanded, more inclusive agroecology. In this text, the words 'farmer' and 'landuser' are synonymous. This inference is more with the word 'landuser' as foresters and agroforesters are also land-use decision makers and, therefore, 'farmers' of the land. In carrying this further, one might employ the expression 'applied or land-use agroecologist'.

Likewise with the term 'farm'. Agriculture, as well as treecrop and tree plantations, are land-use activities that fall squarely under an agroecology heading. All are farms in the broad sense. Pushing the language, there is the term 'agroecologically-productive entity'. Rather than deploying a new phrase, the term 'farm', in its broadest sense, is employed in this text.

In apex agroecology, agriculture can be regarded as a specific subset within agroecology. What is happening is that, as words, agriculture and agroecology will be synonymous. Until this occurs, the narrower, non-agroecological definition of agriculture applies within this text.

For clarity, the word 'agronomy', i.e., the science of crop production also, the management of farm lands (again broadly defined) does not carry a negative connotation, i.e., as with agro-chemical usage. Also, agriculture generally applies to land use without the forestry and agroforestry inclusions. Being broader and connotation free, agronomy is often utilized in place of agriculture.

In a farm situation, economics are not just financial goals. There are a range of objectives. This include the risk profile and the economic orientation. Economic goals can even include an expected environmental outcome. Profit means the difference between revenues and costs.

For lack of a better term, agrotechnology refers to set agroecosystems with specific ecological dynamics, set assemblages of species, and with certain economic expectations. Agrosystems have a broader meaning, agrotechnologies are more precise.

It would be nice if the agrotechnologies were set-in-place systems lacking in variation. This is not the case. These can be modified for specific purpose. This is accomplished by way of the management options and vectors. The options are minor changes that affect, in lessor ways, the economic outcome (not only in profitability, but through the other system objectives).

The stronger term 'vector' implies the capacity to change, not only the economic objectives, but the use situation. Further complicating any outcome, a major change in economic direction can be the result of seemingly minor adjustment in an input or how the system is managed.

Back Topics

Agroecology and agroecological economics are broad topics not lacking in variety and nuances. Not all are covered here, nor is there intent to do so.

This text looks at the more common threads and takes these to their end points. If one wants to study a fuller range of concepts, ideas, principles, practices, and approaches, those that predate this text, the book *Agroecological Economics, Sustainability and Biodiversity* (2008) is suggested. This is by this same author.

Acknowledgment

Although this book did not benefit from institutional support, the majority of the photos were taken while on assignments through the USAID Farmer to Farmer program.

Contents

1

Introduction

Chapter Preview

This chapter sets the conceptional stage including the need for a strong theoretical base. Briefly, need is greatest when data, in sufficient quantity, is in short supply, experiments or trials are costly, or when the information sought in enmeshed in, and hard to extract from, the overall complexity. All these conditions apply to and hinder agroecology and, by extension, to biodiverse-based agronomy, forestry, and agroforestry.

Theory does peak. An apex is reached when the complexity is condensed into base equations or algorithms. The argument in this chapter, and throughout this book, is that this is within reach.

Chapter Contents

Introduction

When this text was published, a debate was raging on how the food, fiber, fuel, and other farm and forest products, those needed for growing populations, were to be obtained. Some were convinced that the world could only be fed through high-input, high-output agriculture. Others disagreed, arguing that sufficient outputs can be obtained from environmentally-friendly, agroecologically-endowed farming.

Despite the argument that high-output agriculture is needed to feed the world, there is opposition. The opposition does not focus solely on genetically-modified (GM) crops, nor on the large volumes of toxic agrochemicals, nor on having huge tracts of a single crop species. It is the entire picture that is worrisome.

Agriculture is the foremost land-use activity, it impacts 50% of the terrestrial surface of the earth (news.nationalgeographic.com, 2005). In an age of climate change, environmental degradation, and species extinctions, agriculture should not be contributory to ever-widening, globally-defined environmental problems. The mere fact that high-output agriculture is cause for concern is indicative of a severe misdirection.

Rather than being an environmental negative, farm landscapes should be welcoming to natural flora and fauna, be more accommodating with regard to quality-of-life issues (pure water, clear air, temperature moderation, scenic landscapes, etc.), and should embrace a range of environmental, economic, and social objectives. Since conventional agriculture is falling short in these wider goals, many seek an alternative.

In agroecology, the ideals represented must be put into practice. To gain acceptance, there must be convincing evidence that the wherewithal exists to produce food for the many. Likewise, it must be demonstrated that the science is capable of fulfilling all the stated objectives. This means answering the technical questions and presenting a clear counter direction.

The New Paradigm

In offering a counter direction, old divisions fade, new facets emerge. Although a lot carries over, there are those concepts, ideas, principles, practices, and approaches that are unique to agroecology. Through these, agroecology blossoms into a new paradigm.

Reordering

In the past, ecology, agriculture, forestry, and agroforesty have been considered separate sciences. The ascent of agroecology brings on a different ordination. The three land-use sciences (agriculture, forestry, and agroforestry) become subsets of agroecology. The borders of agriculture and forestry become blurred to the point where merger is a distant, and distinct, possibility.

This fusion is a given. Trees can be as ecologically causative in farm settings as they are in natural ecosystems. Agroforestry and forestry, applied in an agronomic landscape, can provide an economic and ecological balance. Moreover, as land-use sciences, the three subsets share the same theoretical core.

Within this reordering, agroecology, along with natural ecology, are the main subdivisions of ecology. Although there are a number of factors involved in this reordering, this is natural outgrowth of the rise of ecology, the desire to incorporate more biodiversity in agriculture, and the budding theoretical base of agroecology.

The dividing line between natural ecology and agroecology is distinct. Hunting and gathering in natural ecosystems is natural ecology. If the hunting and gathering involve ecosystem modification, i.e., to increase the available catch or harvest, this crosses the line into agroecology.

Photo 1.1. A mix of banana and papaya. This, and the subsequent photo, are of commercially-formulated intercrops. These contrast with more ad hoc, subsistence systems that are often smaller in area, singular in design, and harder to quantify. Both applications promises higher, per-area total yields, as well as reduced risk.

There is another dividing line. By extension, vegetative modifications have an economic goal. It is the application of economic criteria that differentiates agroecology from natural ecology.

Differences

The implications of a reordered and expanded agroecology, as an ecological science, are significant. Important is the shift away from the green revolution model.

The green revolution model is defined as high-yielding monocultures supported almost entirely by outside, mostly agro-chemical, inputs. Genetically modified (GM) crops are often found. (Agro)ecology is generally not part of this picture.

Just as the logical extension of the green revolution model is the very large-scale monoculture, i.e., the industrial or factory farm, the logical endpoint of agroecology is an agronomic landscape that is very biodiverse and very ecologically active. Clearly, these can be diametric in view and expression.

These are not the only opposites with the new paradigm. As separate sciences, agriculture, forestry and, more recently, agroforestry place an almost complete reliance upon empiricism.

There may be disagreement on the role empiricism plays. The notion here is that ecology, and by extension agroecology, represent a far richer theoretical base. By comparison, the theory behind traditional agriculture is fairly insignificant.

Without the abstract insight provided by an agroecological base, agriculture has been confined to experimentally simple paths. The requirement for experimental simplicity has sidelined more bio-complex approaches and has biased the knowledge base in favor of a mono-crop supported by single-purpose inputs (insecticides, herbicides, etc.).

In contrast, agroecology can tackle, by way of theory (as exhibited throughout this text), the different expressions of biodiversity. This is not the only gain.

Divergences

Agricultural alternatives come by way of competing titles.

- Some are residual from earlier efforts, e.g., pomology;
- some have long histories, but new names, e.g., intercropping and agroforestry;

- some are based on imposed limitations, e.g., organic agriculture;
- some focus on a range of practices or approaches, e.g., conservation agriculture;
- some express a philosophy and, through this, have gained prominence, e.g., permiculture;
- some exist to exploit nature-supplied efficiencies, e.g., synergistic and forest gardening (agroforests);
- others, some being less noticed, focus on one aspect or practice, e.g., hügelkultur and waru waru.

Shorn of their philosophies, these alternatives represent overlapping subdivisions in a vast continuum. The common denominator lies in their inherent ecology. Across this continuum, the theoretical base of agroecology envelops all the divergent titles.

New Applications/New Directions

Agroecology brings expanded scope and cognition. From this comes various manifestations of biodiversity. These can play a greater, if not a major role, in food and fiber production. Examples are numerous.

For a two-species intercrop, there are thousands of crop species that can be grown together to economic and environmental advantage. Add variables, such as spacings and planting densities, and the number of permutations may reach into the millions.

Despite yields that far exceed those from high-output monocrops. comparatively few intercrop combinations have been researched. The demand for field trials could lessen with a pre-trial sorting of the possibilities. Only those with the most potential would be advanced.

Most of the intercrops fall into an economic convention, that of increased yields. There are the atypical intercrops, those that can increase outputs while decreasing costs. These often peak interest. Once identified, these multi-task intercrops could have a prominent role.

Going further into bio-complexity, the advantages multiply. Throughout the world, there are many extensive-in-area wood or treecrop plantations. This latter include oil palm and rubber. It is more than possible to retool these such that, through added biodiversity, they produce a variety of foodstuffs and other useful outputs. This need not diminish their primary purpose.

This plunge into world of agro-biodiversity has its obstacles. There are the often overwhelming number of options and the implementation details. This latter can keep a system from its intended purpose. Extracting the full use benefits from agroecology is, and will be, a daunting task (De Schutler, 2012).

Adding to the problems, there is little data and analytical insight that would facilitate implementation of the many known expressions of biodiversity. There are also some potentially useful, but undocumented, formulations of biodiversity that have yet to receive mention in either the scientific or the popular literature.

Due to the shear magnitude of what is needed, data in sufficient quantities will not be available in the foreseeable, nor the distant future. Rather than wait, it may be better to proceed on with what is known. Theoretical analysis can go far in addressing many of the knowledge gaps.

Agroecological Theory

With the exception of those involving land use, all sciences, including ecology, have and continue to benefit from abstract theory. This is key aspect of science. This has profound effects as theory (a) increases understanding, (b) helps reveal the full range of options, (c) focuses and streamlines the research process, and (d) builds future capacity.

Understanding

In remote corners of the world, farmers have found ways to produce crops with limited resources, often under trying climatic conditions. Likewise, through the history of western agriculture, similar climate-adopted practices have been developed.

As agriculture simplified, these time-tested methods have been sidelined. If reinstated, some of these offer high yields and, at times, at a lower cost. Some also offer a better distribution of labor or can reduce climatic and other forms of risk. As these systems can be quite sophisticated, a theoretical unraveling helps in understanding and, ultimately, in finding applications in ecologically-balanced, productive landscapes.

Options Revealed

It is important that a theoretical core lead to on-farm benefits. This is done through systematic listings and detailed explanations of the ecological tools and options that farm and forest researchers have at their disposal. One facet lies in establishing the links between the different expressions of biodiversity and the varying farm and/or forest objectives.

A prominent example is the theory behind agricultural disarray, i.e., the lack of clear, in-field spacial patterns. Disarray is common in parts of Africa. When mentioned, it was often assumed a negative, not an evolved technique to counter climatic uncertainty.

Photo 1.2. A more biodiverse, commercially-applied intercrop. This features a mix of banana, lime, cassava, and cocoa. Both this and the proceeding photo were taken in the Dominican Republic.

There are other options that have been overlooked. For centuries, farmers relied upon free-ranging chickens, geese, ducks, and other domestic fowl to control crop-eating insects. When deployed in combination with insect-consuming, natural bird species, this becomes ecologically-elegant solution to a common problem. Although almost completely neglected, theoretical explanation sets the stage for research into this old, but ecologically promising, counter (as detailed on page 208).

Research Advanced

Because of the specific needs of farmers and foresters, the rich theoretical base of natural ecology is not always applicable. A lot of this is due to the economic divergence.

Agroecology has trailed ecology in theoretical development. The situation has improved. Those sections of ecological theory, those that have direct application, have been expanded. Also, by compiling, reconciling, and augmenting the bits of agro-specific theory scattered throughout the literature, a core has been established. This has included adding in missing concepts and previously overlooked areas of study. An example of the latter is landscape agroecology.

There are two aspects to theory; breath and depth. The first involves capacity building, looking at the full range of possibilities and, as with the chicken/bird example, fills the knowledge gaps. The second is depth, pushing theory to its limits. The latter may not have direct on-farm benefits, but it does aid the research process and this eventually carries into the applied.

Capacity Building

Theory influences, directly or indirectly, all aspects of agroecology. The benefits would be global with application potential in food security, climate change, and in addressing a wide range of environmental concerns. There are some immediate gains.

The literature base of agriculture and forestry is overwhelmingly large. With much to be extracted from records stretching back into the distant past, comprehensive theory provides a framework from which the most contributory studies can be categorized and incorporated into the new agroecological paradigm.

Carrying this forward, the research in agroecology and agro-biodiversity is far less, more scattered, and does not always speak to the whole. This means that large gaps exist in what should be an uninterrupted knowledge continuum. There is a need to leverage existing studies and observations, those that deal with specific situations, into a larger understanding. Instead of standing alone, this are used to construct or validate models, strengthen existing theories and, in combination, add new theoretical layers.

As theory advances, gaps are filled and the mosaic closes. As this comes about, it becomes possible to layer on broader, more inclusive explanations. This implies the existence of an uppermost layer or layers where theory distills into a few basic concepts and equations. Although speculative, this is part of an understanding that will eventually project and define agroecology.

The Theoretical Core

This text pushes the limits of theoretical agroecology. The notion is that theory must proceed practice. It also important that theory portend practice. This is done by looking at the far recesses and the unexplored upper levels.

In doing the latter, this text enters the rarefied air of high theory. The air is rarefied by the lack of data. There are steps, other than redoubled data collection, to rectify this.

The starting point is abstract theory. These are the less precise, verbally-described interrelationships that underlie (agro)ecological dynamics.

Abstract theory gives way to more specific, better understood, qualified theory. This is where equations have been proposed, but lack the numbers to delineate their actual form. Unquantified, mathematical expression is more definitive and denotes a greater level of understanding than verbalized descriptions.

The best case is quantified theory. Employing theoretically-sound relationships, equations have been derived. This are set, either through statistical means or through an analytic inference. Beyond this, the equations need to be site, crop, and situation calibrated. This requires far less data and this helps unrarefy the air.

For many, entering the rarefied air of high theory may seem daunting, even unnecessary. It does serve a larger purpose.

Instead of a transition to agroecology by way of agroecology light (a slow drift from mono-agriculture and the green revolution model to basic biodiversity), deeper, wider, and firmer theoretical underpinnings allow for a quicker immersion into full-fledged, option-rich agroecology. In many cases, this is where the full gains reside.

Apex Analytical Algorithms

Theory is not endless. A point is reached where the central elements become visible through the informational clutter, equations distill, and a conceptional apex is reached. This is agroecology condensed into core concepts and theorems. These describe how crops relate to each other, the land, and their surroundings.

Agroecology is economically based. These underpinnings come to the fore as theory peaks. It is from the core theorems that the evaluation algorithms and the economic methodologies are derived. The aim of this text is to advance the core concepts and theorems and, from these, to present accompanying economic analytics.

In doing so, it must be acknowledged that there are practical limits. Most, but not all, situations can be condensed into a few quantitative statements. Agroecology may be far too complex for a fully encompassing set of equations. Those analytical algorithms that exist are starting points. The finer details and the intangibles, e.g., societal needs and norms, are added to the conceptional frameworks put forth in this text.

2

Core Concepts

Chapter Preview

This chapter, and the next two, set the stage for deriving the base equations. First comes a system of classification. Based around recognized agrotechnologies, this is a requirement. The removal of plant species from the classification shifts the focus to the ecology inherent in each agrosystem. Examples illustrate how agrotechnologies functions as a first step condensate.

Chapter Contents

Vector Categories

- Agrobiodiversity
- Facilitative Biodiversity
- Genetic Improvement
- Varietal
- Microbes
- Rotations
- Fire
- Landscape
- Location
- Physical Land Modifications
- Ex-Plot/Ex-Farm Inputs
- Environmental Setting/Management Inputs

Introduction

Agroecology, being a broad science, offers a multitude of approaches to produce food, fuel, fiber, and other products destined for human consumption. When dealing with individual farm or forestry plots, there is a need to determine when one agronomic approach or an agroecosystem application is truly different from another. Part of these is in finding, among the many possibilities, those approaches that offer the most.

Ideals

Agroecology is based upon a series of ideals. There is the notion that nature and natural ecology should be the basis for any and all agronomic activity. Toward this end, natural solutions should be able to resolve the full range of cropping problems. These include improving soils, controlling herbivore insects, reducing plant diseases, and eliminating weeds.

Idealistic as this might seem, this is not unattainable. There are agroecosystems that come very close to eradicating the full range of problems. Notably, these are the highly biodiverse agroforests which, through their composition and layout, are almost entirely nature managed.

When high levels of biodiversity are not possible, the scope for utilizing what nature provides diminishes, but is not entirely lost. The issue is to maintain, as biodiversity levels diminish, as much ecology as possible. This means identifying the tools and approaches and finding how these are best used. This is an analytical question.

Analytical Tools

The first step in implementing nature-dominated or nature-assisted agriculture is to identify the basic tools. A breakdown by the traditional land-use sciences, i.e., agriculture, forestry, and agroforestry, is not all that insightful. The same holds true when looking at those broad classifications that have similar outlooks and goals. Permiculture, organic, sustainable, and low-input agriculture have political and social overtones, but lack the scope and technical precision that comes from being a hard science.

Harnessing natural forces for agronomic gain is not easily expressed as a single topic or through one system of classification. At the lowest level, there is the crop. Maize, rice, wheat, potatoes, etc., are all staples in one region or another and roughly describe some local farming practices. This is fine when there is only one type of agroecosystem. Crop type fails as a descriptive term in more biodiverse agronomic settings.

When expounding specific land uses, there are two systems of categorization. The first of these are the agrotechnologies, the second are the ingrained management options/decision vectors.

The Agrotechnologies

Given all the feasible agroecosystem design variations, i.e., species combinations, planting ratios, spatial and temporal patterns, etc., the number of possibilities is extremely large. To shortcut the process, key points or regions along the vast continuum of possibilities (Figure 2.1) have been identified. These are the agrotechnologies.

Agrotechnologies are discrete agrosystems containing one or more plant species where, through unique ecological dynamics, each has as a specific purpose and a singular productive, economic, and environmental profile.

Each of the identified agrotechnologies has the potential for a positive contribution to a farm landscape. Each does so in a distinct way. Some are utilized directly in the form and/or design first identified, others may require slight modification for a better ecological and/or economic fit.

Agrotechnologies also include some non-biodiverse expressions. These are those land modifications, i.e., irrigation canals, terraces, and similar structures, that increase the productive potential of a nearby or even a distant crop plot.

In addition, there are some purely facilitative agrotechnologies that, although they produce little or nothing, do contribute ecology and the economics of farm or forest landscapes. These are the bio-structures, i.e., windbreaks, hedge fencing, riparian buffers, etc.

Categories

As points along a multidimensional continuum, no system of classification is free of overlaps. Without a clear winner, agrotechnologies can be categorized in any number of ways. These can be sorted by their underlying economics. Risk is a decision factor and the different types of uncertainty, i.e., climate, insect, disease, etc., can be a basis for grouping agrosystems. Another categorization criteria is environmental compatibility, i.e., how well an agrosystem supports local flora and fauna.

Subdivided along the lines of traditional disciplines, the broad groupings are

(1) Productive or principal-mode (including facilitative systems)
   (a) agronomic
   (b) agroforestry
   (c) silvicultural
(2) Temporal
(3) Auxiliary systems
   (a) land modifications
   (b) landscape bio-structures

Some categorizations follow biodiversity lines. This might be

(1) Productive (agrobiodiversity)
   (a) monocultures
   (b) intercropping
   (c) fauna based (including pastures)
   (d) complex agrosystems
(2) Facilitative (biodiversity)
(3) Temporal
   (a) fallows
   (b) rotations
   (c) taungyas
   (d) continual
(4) Auxiliary Systems
   (a) land modifications
   (b) landscape bio-structures

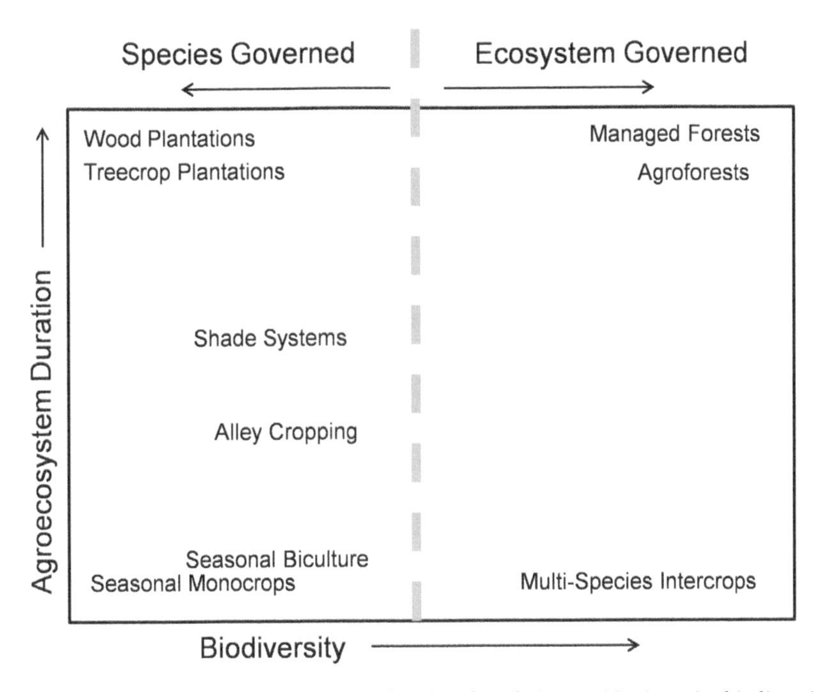

Figure 2.1. The agroecological continuum showing the relative positioning, visa biodiversity and agrosystem longevity, of key agrotechnologies. This surface can also be subdivided into left and right, the left being the plant-governed agrosystems. The right side are those that are ecosystem governed.

Explained

The agrotechnological system of classification starts in two dimensions. This is illustrated by way of the agroecological continuum. Two dimensions of this multi-dimensional continuum are shown in Figure 2.1.

In Figure 2.1, the anchor points are key agrotechnologies. These are area labeled. These are the commonly recognized, biodiversity-based agrosystems. In many cases, these subdivide. For example shade systems can be categorized into light and heavy shade (described in this chapter).

Greater insight is gained by adding more dimensions. One of these is the temporal dimension. Others involve management options. As the latter provide direction and magnitude (often economically stated), these are also decision vectors. Listed in this chapter, these vary in strength and importance and define other areas in the now multi-dimensional continuum. As the auxiliary systems can be biodiversity-offering additions, these fall under the same categorization.

At the floor of this system of classification are the species (one or more), spacings, and spacial patterns. In review, some common spatial patterns are pictured in Figure 2.2.

It is possible to use agronomy, agroforestry, and silviculture as top categories. Some may consider this as an anachronistic carry over from a now superseded past. As land-use categories, the traditional disciplines have value and, as a means to pare the study possibilities, they still have functional value.

In Figure 2.1, wood plantations and managed forests are part of silviculture, alley cropping, shade systems, and agroforests are inclusive in agroforestry. The remainder are agronomic agrosystems. If dropped, the agrotechnology is the first subdivision.

As an added note, the agroecological continuum is subdivided into two parts. The center line separates those agrosystems that are species governed (left side) from those that are ecosystem governed (right).

Briefly, species governance is where the individual plants dictate the plot ecology. Ecosystem governance is where, because of exceeded, base-level biodiversity, a plot becomes more than the sum of the plant-dictated ecology. Chapter 11 focuses on those equations that underlie ecosystem governance.

Key Agrotechnologies

A complete list of agrotechnologies is found in Appendix 3, descriptions follow in Appendix 4. This chapter gives a brief description of the secondary and other, commonly-found agrotechnologies. These start with the monoculture.

Monocultures

This is an agroecosystem where, by intent, all plants are of the same species. In nature, plant monocultures are a seldom-seen exception. One example is kunai grass in the highlands of Papua New Guinea.

In the land-use sciences, the monoculture is not the exception. It is the common method to raise most crops.

There are clear reasons. The grouping of like species is advantageous when (1) one crop is substantially more valuable or desirable than others, (2) when a crop is mechanically planted and/or harvested, and/or (3) when the technical knowledge is lacking on how to best plant and manage bio-

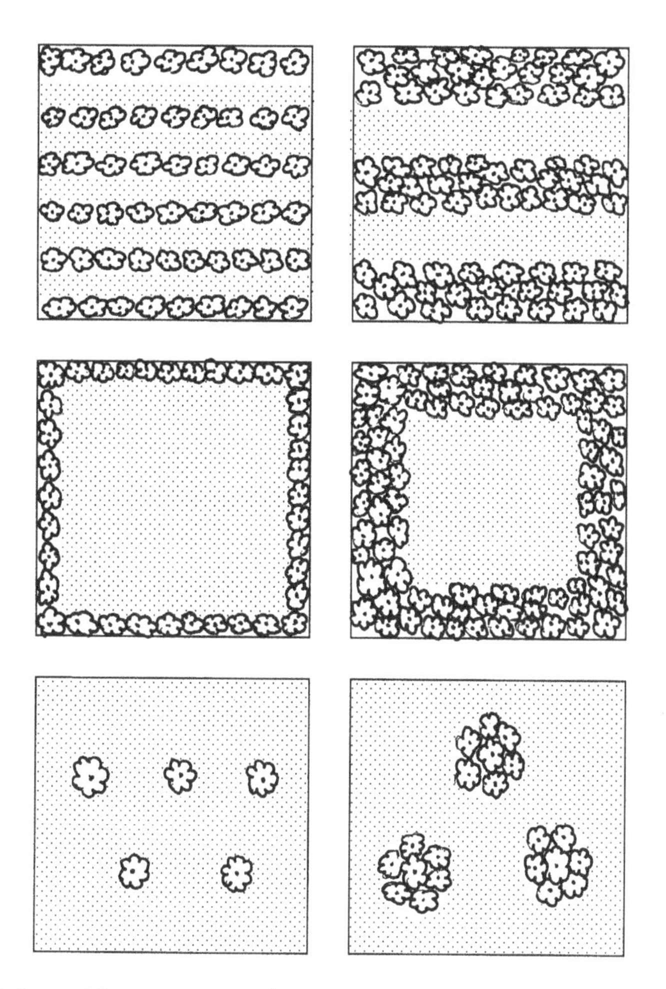

Figure 2.2. Some of the common spacial patterns. On the left, from top to bottom, these are row, boundary, and individual. The right side shows the same as clumps or groups.

complex intercropping alternatives. Monocropping can be seasonal or, as with forest-tree plantations, last many decades.

Intercropping

Another key agrotechnologies is seasonal or multi-seasonal intercropping. This is with mixed species, either as annuals or a mix of non-woody perennials and seasonal crops. There are other intercropping categories. These can involve facilitative species grown to benefit an annual crop, a perennial crop or some woody productive species.

Photo 2.1. A multi-species, seasonally-based intercrop. This examples those systems found in the lower right portion of the agroecological continuum (Figure 2.1).

The majority of examples are two species agrosystems. The agroecological continuum (Figure 2.1, lower right) and Photo 2.1 shows the complex, seasonal intercrop.

Shade Systems

There are situations where an understory species is shaded by a taller-growing companion species. The defining attribute is that the overstory canopies are touching or are in close proximity. The overstory and understory components can be spatially ordered or disarrayed.

As a growth strategy, there are a lot of variations off this theme. In seasonal intercropping, a taller species will over top and take light from other plants. More common are those crops that grow well in shaded environments. The often-cited example is shade-grown coffee. Other variations, those with different overstory tree species, have less shade and might well have a greater diversity in productive species.

Given the ecological and economic dynamics, these neatly divide into heavy and light shade. The economic rational for this presented in Chapter 3 (see Photos 3.2 and 3.3, pages 40 and 41).

Alley Cropping

The planting of crops between tree rows is alley cropping. The basic layout has multiple rows of seasonal crops planted between parallel rows of woody perennials.

To keep light competition from becoming a negative influence, the woody component is pruning to a low height before crops are planted. Pruning also diminishes the amount of essential resources during the competitively critical cropping phase. There are other modifications, these include taller trees where crops are grown in a more pronounced between-tree alley. These are considered facilitative agroforestry systems.

Photo 2.2. A homegarden that surrounds a traditional Ethiopian highlands house. As a type of agroforest, these ad hoc systems are formulated around high levels of agrobiodiversity. Except for species content, these are very similar to the homegardens in other parts of the world (for comparison, see Photo 11.1, page 155).

Agroforests

Not to be confused with the broader agroforestry term, agroforests are species-complex agroecosystems containing a large percentage of woody perennials. Their hallmark is their high levels of biodiversity coupled with spatial disarray.

Variations off this theme are noted for their extremely low levels of inputs and their environmental friendless. The productive role of agroforests is unique. These generally find use, and they are at their best, when they produce many outputs. This, along with their environmental compatibility, dictates their applications and their landscape location.

Their advantages have not gone unnoticed. Agroforests are found in all humid tropical regions. Outside of this, this are occasioned in dryer or in temperate regions.

In terms of their biodiversity levels, these far from the single-species norm. As such, these are more a curiosity, not much is known on their design and how this influences their internal dynamics. Although not studied and analyzed in any depth, these could find use with treecrop plantations (as explored in Chapter 11).

Fallows

The concept of an agroecosystem, one that produces little, but adds to the ecology and the productive potential of a farm, is demonstrated by the fallow. These exist as both as a facilitative and a temporal agrotechnology.

Fallows are a rest period for the land, allowing nutrient recovery and a resetting of the ecological dynamics. The crop-free interval can range from one season every few years to extended fallows lasting decades. The latter are often found in slash-and-burn agriculture. With regard to content, more biodiversity is better.

Rather than being a constant, i.e., with a set period, they are more effective if well formulated for the cropping task at hand (Smith et al., 2008). This might include managing the biodiversity content for the upcoming crops. This is a mostly unstudied topic.

Taungyas

The taungya is a temporal agrotechnology that can be used whenever wood, fruit, or other productive trees are planted. The ecology of the taungya is based on utilizing, for economic gain, the excess light, water, and other essential resources during the first few years when the trees are small.

This allows for seasonal or short-duration crops in an orchard, treecrop plantation, or forest-tree plantation.

There are variations off this temporal theme. Besides being categorized as temporal agrotechnologies, these are also, from a non-temporal perspective, intercropping or shade systems.

Infiltration Barriers

There are different methods to prevent erosion and to allow for water infiltration. One form is the barrier. These are contour or cross-contour ditches, bunds, and/or hedges that impede the water flow down slopes and hillsides.

In their more intense form, infiltration barriers are employed with seasonal crops and where soil is easily eroded. In a less intense form, these are found with perennial crops, As agrotechnologies, infiltration systems keep water within crop plots or direct such flows into the plot. Generally, these are considered separate agrotechnologies and not part of the productive agrotechnology being supported.

Live Fencing

Fencing can be more than the partitioning of farms and farm landscapes. This can be, if the right type of fencing is utilized, an agroecologically-positive landscape addition.

Living fences are single rows of vegetation, usually shrubs or pruned trees, that contain animals and demarcate land. In addition to their role as a fence, these can serve as infiltration structures, riparian buffers, for erosion control, as corridors for the movement of insect-eating insects, and have lessor purpose as windbreaks.

These are classified as landscape bio-structures. Normally, these offer no direct output.

Management Options/Decision Vectors

Agrotechnologies are not set, unchanging systems. There are those variables (including inputs, system-internal design alternatives, and landscape factors) that, individually or in combination, add or subtract from the core ecology. Under the agrotechnology system of classification, these are dimensions in the roughly presented agroecological continuum.

Photo 2.3. Live fencing showing wire supported by planted trees. This design provides animal, flood, and wind protection for an enclosed banana plantation. The photo is from Dominican Republic.

As management options, they are just that, they change the agrotechnologies in minor or in major ways. These are changed in response user objectives.

Some of these management-induced changes are tasked with improving the economics. These have little impact on the core ecology and the agrotechnological intent. There are a few that, seemingly insignificant, impart a major change in the ecology, economics, and overall outlook. These are the decision vectors.

As an example, a different planting method in a forest-tree plantation can seriously reduce the rotational time. The improved economic outcome financially justifies a better site (e.g., prime farmland), more inputs, and more intense management. In-turn, these improvements opens the possibilities for adding productive biodiversity in the form of understory crops. There is more on this example at the end of Chapter 9.

The options/vectors are not an exact science. Some study is required to determine which of the options are best employed and in what situations they can be most effective. A more difficult task is the identification of those that impart major change in the ecology and economics and rise to the level of directional vector. These are analytical issues.

Vector Categories

Vector theory is the study of the strengths and potentials of the various options to achieve or augment the goals of a base agrotechnology. The list of options/vectors reads as

agrobiodiversity
facilitative biodiversity
genetic improvement
varietal
microbial
rotations (with or without a fallow)
fire
landscape
location
physical land modifications
ex-plot/ex-farm inputs
environmental setting

Remembering that, within each plot, there exists species (one or more), spacings, and spacial patterns. These, coupled with the management options, described below, offer innumerable combinations.

Agrobiodiversity

The productive intercrop, through increased biodiversity, puts a face on agroecology. The productive intercrop is where multiple species, each offering economically-viable yields, are mixed and ecologically interact within a farm plot.

With agrobiodiversity, the emphasis is only on planned for additions, non-producing species, weeds included, are often not considered. Within this context, there are varying forms of intended biodiversity. To this, more can be added.

The first and strongest form is archetype agrobiodiversity. This is where each species is integral in the planned economic outcome. The common case is intercropping where maize, bean, and squash provide mutual benefit and three harvestable crops. These relationships ecologically define a core or base agrotechnology.

The next situation is expanded agrobiodiversity. With this situation, minor species substitution can take place without changing the ecological character or economic emphasis of the planned system. This might occur where multi-species are inserted in place of single species, e.g., growing multiple species of maize and/or beans in the bean/maize intercrop.

Under agro-enrichment, unused space, unexploited resources, and/or an unused space is exploited. This can occur with complex agroecosystem where light strikes the ground. Under the rules of such systems, this means positioning another plant. In the spirit of biodiversity, often another species is established.

With casual agrobiodiversity, an established perennial of a new addition will interfere and change the anticipated outcome. Because the intrusion consists of a valued output, it remains. The common case is a fruit tree that remains from a previous land use or one that spontaneously spouts.

Yet another application is supplementary biodiversity. This is where a productive species can be added without subtracting from or altering the overall ecosystem. A low density vine may be added above forest or treecrop plantation. One case has rattan growing atop rubber trees. The low density and occasional and destructive harvest of the vine has little or no impact on rubber production.

With all types of agrobiodiversity, there is a general requirement for rich soils. Along these same lines, the nutrients being removed through harvests should balance those that enter a site, either though natural means or those that are externally applied.

Facilitative Biodiversity

On any one site, a crop species might not grow well but, with association with another species, the situation could change. This is when one species facilitates the growth of another. These associations can integral and planned part of the base agrotechnology.

Facilitation is where the presence of one plant species benefits to another. The example are numerous. Facilitation increase amount of water and essential minerals in the soil. The hope is that this will increased yields. This form of facilitation is more associated with the base agrotechnology.

When facilitative biodiversity is added to an existing agrotechnology, it is more to guard productive plants from climate related threats or to protect these same species unwanted and uninvited flora and fauna, e.g., weeds, herbivore insects, grain-eating birds, etc.

There are other cases where facilitative biodiversity plays a different role. Deep rooted plants can be added to interlock roots with and protect orchard trees from toppling.

There is also casual biodiversity. Without being invited, plants often colonize a site. Many can play a facilitative role, e.g., in harboring beneficial insects, provided they are compatible with the primary species (one or more). These are allowed to stay if their value exceeds any losses incurred.

Genetic Improvement

The need to genetically improve crops has been a force from the inception of agriculture. It is safe to say that fruits and vegetables that were raised hundreds, if not thousands of years ago, would be slightly different, or all-together different, than their modern equivalents.

Until comparatively recently, this was an informal process where species where improved by replanting those with the most desirable characteristics. Wheat became a cultivable, stable crop after wild varieties were found that did not shed their seeds upon maturity.

In another part of the world, maize underwent a number of transitions. The first was the jump from unassuming perennial plant, teosinte, to something of interest to early mezzo-American farmers. Through many centuries, informal selection continued and brought maize from its sub-tropical origins to a major crop in tropical and temperate regions.

Not all gains are as momentous. The seedless banana was major step in becoming a universal crop (the seeded-bearing fruits being challenging to eat). Seedless watermelons and oranges were less world-shattering, but still represented genetic progress.

As mention, a genetic change, specifically high-yielding staple crops, in combination with monoculture and chemical inputs led the green revolution model. Other agrotechnologies can benefit from changes in the genetic characteristics of their component species. The most obvious attribute is improved shade resistance.

For many crops, multiple varieties exist. These can number in the hundreds. A past tally for wheat found over 600 species and varieties (Percival, 1922). As an under-utilized resource, most of those that have fallen from use. Even with the established varieties, their response to climatic, soil, moisture and other variables is not well documented. This is an impediment to optimized use.

Despite these unknowns, regional varieties can be the better choice as some, or many, have been adopted to and time tested for local climates and local soil. They also may be best at resisting localized threats. As an added advantage, these varieties can find favor and command higher prices in regional markets. In general, finding the right crop and the right variety may seem intuitive, but this remains an underexploited possibility.

Varietal

The notion of huge field of a single-crop variety is one of the missteps of the green revolution approach. Without going into full biodiversity, one step in this direction is the multi-varietal monoculture. With enough intra-species genetic variation in amongst varieties of the same species, the agrosystem can have some of the ecologically advantages of a field of different crop types.

The varietal vector could have greater application wood-producing, forest-tree plantations. An example might lie with pines or similar species where processors and consumers are less likely to notice the change.

Microbes

In nature, microbes serve any number of purposes. Microbes have proven to be both a harm and a blessing. If developed to accentuate the positive, microbes have been reported to accomplish many useful agronomic tasks. Well know is the nitrogen-fixing mycorrhizae associated with pine trees. If not inoculated with these microbes, poor soils can equate with poor growth.

There are a host of other useful tasks associated with microbes, e.g., Wikipedia—Microbial Inoculant (2013): Gaiud (2011); de Vrieze (2015). This includes breaking down chemical compounds to release nutrients to plants, e.g., phosphate-solubilitizing bacteria. Microbes also decompose vegetative residues left after cropping and, through increased in-soil carbon, can improve water availability. They can also add resistance to common plant diseases.

Endophytes (in-plant living fungi) help plants withstand higher temperatures than would be possible. Other endophytes confer drought resistance. Some may even help maintain yields when, as within an intercrop, plants must grown in less than full sunlight.

Rotations

The notion of different crops grown on the same plot of land in different seasons has a long history. This is a proven means to improve soil nutrients, to combat some insect species, to downgrade plant diseases, and to keep weeds at bay.

As a management vector, rotations would be an addition where seasonal agrotechnologies are strengthened through rotations. A key reason for rotations is nutritional, using one planting to prepare the nutritional setting for an upcoming crop. This, plus any mineral that are added between seasons, should present a close-to or an ideal balance of nutrients for an incoming crop. These have a negative impact on diseases and some detrimental insects.

In simplest form, simple rotations are addition to cropping system and could be classified as vector. A more elaborate form, that of the taungya, is listed as an agrotechnology. This is because these can have a one, multi-decade, primary species. Biodiversity is added, and removed, as the system progresses.

Fire

The use of fire within agriculture and silviculture is a specialized topic. There are times and places of use. In other situations, fire is a danger to be avoided.

Fire can be the key element in a slash and burn strategy. The burn release nutrients for the immediate use of the crop. An intense burn also kills in-soil weed seeds and in-soil insect eggs and grubs.

To the negative, burning strips the soil of any protective layer, making it more susceptible to erosion. Through fire, a large percentage of the in-biomass nutrients are lost to the system.

The alternative, natural decay, through a ground cover of dead biomass, protects the soil from erosion. The problem is that the contained nutrients are released at a slow rate, often missing the peak need period of the subsequent crop.

In some forest plantations, fire can be, depending on the location and the tree species, a seasonal option. Use is a function of frequency where shorter intervals bring less danger and less risk from unwanted destruction from possible future, more intense fires.

Landscape

The landscape that surrounds a crop plot has spillover effects that, if beneficial, can help increase yields and lower costs. The common example is the windbreak. If well placed, this can have positive effect on crop drying and water use. This allows yields higher than what would be predicted solely though the amount of rainfall.

Other gains are possible. The amount of predator insects is often a function of the vegetative mix in the neighboring plots. As these predator types move into the crop plot, these can provide almost complete protection against an outbreak of herbivore insects.

Landscape gains extend in other directions. Well designed, well formulated landscape can reduce risk and even reduce labor costs. These come about through positioning of the crop plots with regard to roads and other farm structures.

Location

With the advent and use of commercial fertilizers, the seemingly obvious notion of locating crops where they grow best has been sidelined. Although apparent, this is not always followed. In high-input agriculture, the tendency is to add water and nutrients to the soil until suitable yields are forthcoming. A more nuanced approach is to fit the crop and the agrosystem to the soil and surrounding conditions, even if this means subdividing non-uniform plots.

Photos 2.4 and 2.5 are examples of a locational strategy. These show rice being grown in an unmodified stream (Photo 2.4) and a constructed paddy (Photo 2.5).

Biodiversity, especially in limited form, is not a free pass to locate anywhere. Biodiverse agrotechnologies, as well as with mono-crops, should be placed where they perform best and, by extension, where they do the least harm. The implication here are not only with regard to soil fertility, but for factors such as erosion control.

Photo 2.4. Location strategies, this photo shows rice being grown in an active, flowing stream.

Physical Land Modifications

In trying regions and/or with some crops, crop cultivation would be difficult without land modifications. The common changes include adding terraces, paddies, and ditches of various forms and purpose. Lacking land modifications, such as terraces, agriculture on steep hillsides could be an erosion-plagued undertaking. Paddies produce the swamp-like environmental preferred by water-loving rice. Modification of low lands to become rice paddies is an often-cited example (as in Photo 2.5).

Other modifications combat flooding and drought and some are based on changing the structure and usefulness of the soil. The latter is the addition of charcoal (biochar) to improve the nutrient and water-holding capacity. Charcoal is also an ex-plot or ex-farm input.

Ex-Plot/Ex-Farm Inputs

Any outside application to a crop plot would be considered an ex-plot addition. These range from manures to green biomass to agro-chemicals. Another input is water as applied through irrigation.

Photo 2.5. Also locational, but with a land modification, this photo is of lowland, paddy-grown rice. Both photos are from West Africa.

Most of the controversy comes when inputs involve toxic or less toxic chemicals. These inputs include fertilizers, insecticides, herbicides, and fungicides. Although the once highly-toxic agro-chemicals have been mellowed over time, there is still concern. They can still cause unanticipated problems as they drift across time and place, e.g., Strom (2013b).

Less controversial are those that pose miniscule or no health threat to people or to nearby fauna. These are, as insecticides, soap, diatomaceous earth, wood ash, and neem oil and, as a fungicide, milk.

As an indictment of the research community, these greener alternatives, when compared with the riskier commercial chemicals, have far less research backing. Despite their effectiveness, these are seldom, if at all, advertized. To rectify the lack of attention by the science community, some of what is known (part anecdotal) is summarized in Appendix 2.

Among the plot inputs are farm outputs. These can be green manure (cut and carried biomass) or various types of animal or bird manures. These were once the historic norm and provide benefits other than supplying nutrients to crops. Animal manures help in controlling some harmful insects (Morales et al., 2001; Brown and Tworkoski, 2004). They also improve the soil structure and water holding capacity.

Adding biochar (also called tierra prieta), i.e., the mixing of charcoal into the soil, is an input with permanent effects. This increases water and nutrient holding capacity and an improves the bulk density of the soil (Photo 3.1).

Environmental Setting/Management Inputs

The environmental setting is a catch-all category for those practices not included above. This is a collection of agronomically-diverse implementations that, if taken individually. do not constitute a change in approach but, if well chosen, can have a large impact. At lot depends on the overall design and purpose of the agrotechnology being supported. The practices and methods include

- planting (the latter three are for trees)
    - seed
    - seedling
    - stripling
    - stem
    - stump
- pollinating (e.g., adding bees and hives)
- pruning (mostly with tree components)
    - coppicing
    - pollarding

stem pruning
lopping
branch thinning

tilling
traps (as an insect or rodent counter)
weeding (hand methods)

total
select
near plant
in-row

A brief explanation follows in Appendix 2. These have an impact. How the soil is tilled must suit the crop. More intrusive tillage, e.g., with high speed, rotary blades, can grind up useful earthworms. Tillage is also important in weed management. Weeds, being the more immediate threat, often determine the approach.

Environmentally, much the same occurs with high-speed rotary mowers. Along field margins, these can destroy beneficial insects and reduce advantageous predator/prey relationships. A lighter touch may be less costly and less harmful to the favorable micro-fauna residing at the plot margins.

There are other management possibilities. Introduced chickens, in league with natural bird species, if allowed to roam fields could be a potent anti-insect strategy (Wojtkowski, 2004). Although common on European, African, and Asian farms for hundreds, if not thousands, of years, there is no supporting research and no confirmed best methods.

There are other bird approaches, using local species, e.g., wrens and woodpeckers, that function in much the same way. Again, there is no supporting research. To fill this rather large knowledge gap, some information on bird-centered insect control is presented in Appendix 2.

3

Economic Underpinnings

Chapter Preview

Continuing on with a review of core concepts, this chapter looks at what constitute farm goals or, mathematically expressed, the farm and/or landuser objective function. As part of this, there are economic measures that are unique to agroecology. These are the ratios. There are also cost considerations that tend to be missing in bio-simplified revenue-oriented agrosystems. These and other approaches provide a complete platform for economic analysis.

Chapter Contents

Introduction

Because the majority of the decisions are taken along the economic front, any review of core agroecology must include a look at the underlying economics. It is a truism that farms, of whatever size, operate on a series of business decisions. It is always nice when revenues exceed expenses and farm families are well fed.

This often starts with a mix of agroecosystems. This can be different crops within single agrotechnology, e.g., rotations with monocultures of various crop species. More can be done. A single farm can be far more biodiverse with mix of agrotechnologies that employ different combinations of vector variables and agrotechnologies. This may be more the norm.

The complexity multiplies as more options are added. There are those bio-structures that contribute to the ecology and reduce costs but, in of themselves, do not add revenue. The land-modification agrotechnologies are investments where it is good to justify, through revenue gains, the money spent.

These business decisions do not apply equally. Subsistence farmers face many the of same decisions on inputs and yields. Since starvation is a possibility, they have less marginal for crop loss. Therefore, risk reduction has a greater role in the decision process. Whatever the case, business or subsistence, economics provides key insights.

Objectives

As with many endeavors, agroecology often looks at multiple goals. These include

profitability (revenue-costs)
risk reduction
economic orientation
resource (labor, land, and expense) allocation
return on investment
environmental

From these, landusers seek to optimize the use of the land. The base is often conventional economics. This is especially true when the focus is on revenue, risk, return, etc.

Conventional Economics

In common with agriculture, conventional economic measures have clear application. Raw yield figures, along with selling prices and costs of production, indicate profitability. There is no dispute that raw figures, yields, costs, return on investment, etc., can be the basis for economic decisions. The world works around conventional economics and, in making ex-farm comparisons, conventional methods cannot be entirely replaced.

The above-stated objectives can be, and often are, monetarily expressed. The alternative is to utilize a non-currency form. In a departure from conventional economics, non-currency expressions, mainly ratios, dominate this chapter and this text.

Photo 3.1. Sweet potato being grown in charcoal-enriched soil (biochar). This is land modification and an investment that ensures improved yields through improved soil structure.

Ratios

Agroecologists expect more and ecological dynamics plays important role achieving the desired outcome(s). Delving deeper into ecological and economic evaluation requires ratio-bases analysis. As a decision tool, ratios parallel conventional economics and, under some circumstances, the results can deviate from what is projected using conventional economic data.

Ratios have a greater importance than slight differences in outcome would suggest. Ratios go to the heart or essence of agroecology. They provide a degree of ecological insight not attainable through conventional analysis. They also bring up those economic aspects of agroecology that have been overlooked in conventional analysis.

First of all, the advantages of the ratios lie in their ease of use and their intuitiveness. Monetary units cloud the results. Shorn of monetary values, i.e., crop and input prices, these provide a raw and insightful look at the underlying agronomic dynamics. Since all systems are evaluated using the same base, i.e., starting with monocultural yields, this measure is universally comparable. With base value of unity, the numbers calculated are logical, in many cases not needing further interpretation.

As a decision tool, these serve to differentiate the many expressions of biodiversity. Ratios are very useful with models. They are more intuitive and provide a better fit with equations.

All the equations and development in this text are based on ratios. This approach to agroecology and agroecological economics begins with the LER.

Land Equivalent Ratio

A key measure of intercropping success, the Land Equivalent Ratio (LER) is the mainstay of agroecological economics. Proposed by Mead and Willey (1980), this is mathematically expressed as

$$LER = (Y_{ab}/Y_a) + (Y_{ba}/Y_b)$$

where Y_a and Y_b are the monocultural yields of species *a* and *b*. Y_{ab} being yield output of species *a* grown in close association with species *b*. Correspondingly, Y_{ba} is the productivity of species *b* when in combination with species *a*.

Calculation can be demonstrated with a hypothetical example. For the denominators of the above equation, the normal monocultural yield of species a is set at 5,000 kgs per ha while, on the same or a similar site, the monocultural output of species b is 8,000 kgs per ha. When closely intercropped, the yields of the two species, a and b, are respectively 3,000 and 4,800 kgs, the LER calculation will be

$$LER = (3{,}000/5{,}000) + (4{,}800/8{,}000) = 1.2$$

For most crops, seasonal or perennial, annual yields serve well computationally. In a few cases, multi-year averages are needed. For silvicultural purpose, annual tree growth, e.g., the yearly diameter increase, can be substituted for crop yields.

For this, monocultures have a defined value of one. Intercrops with values greater than one, as above, are considered site and cross-species compatible and of greater economic interest than a monoculture. The highest LER values reported exceeds 3.5.

Being a flexible tool, the LER is open to many variations beyond the biculture. The basic equation applies when a productive species is paired with a non-productive, facilitative species, e.g., $LER = (Y_{ab}/Y_a) + (0_{ba})$.

The LER also finds use with systems producing species. In triculture form, LER is expressed as $LER = (Y_{abc}/Y_a) + (Y_{bac}/Y_b) + (Y_{cba}/Y_c)$. This has species c grown together with species a and b. There is no change in the base value (1.0) although one might hope for a higher LER.

Relative Value Total

Some advocate adding prices to the LER. This resulted in the Relative Value Total (RVT). Formulated by Vandermeer (1989), this is

$$RVT = (p_a Y_{ab} + p_b Y_{ba})/p_a Y_a$$

where Y_{ab} is the yield of species a grown in close association with species b, Y_{ba} is the yield of species b raise in conjunction with species a. Y_a is the monoculture yield of species a raised on a soil-similar, same sized plot as the a with b intercrop. The selling price for species a is p_a and, for species b, the selling or market price is p_b.

Used alone, the RVT is of lesser interest than the LER. In most applications, the RVT and the cost equivalent ratio, jointly employed, are more revealing.

Cost Equivalent Ratio

The LER and its derivatives may be the premiere measures, but these do not answer all the economic questions. A fuller picture is obtained through the cost equivalent ratio or though a parallel, economic profit-loss analysis.

Whereas the LER and the RVT put the focus on the productivity and revenue, some agroetchnologies are formulated and achieve success due to their cost efficiency. The Cost Equivalent Ratio (CER) takes input efficiency into account.

The equation is

$$CER = c_a / c_{ab}$$

where c_a is the costs of production for species a, c_{ab} are the costs of production when species a is growth together with species b. An example would be when costs are 20% less when a facilitative species reduces weeding or other inputs. For this, the

$$CER = 1.0/0.8 = 1.25$$

This would be in contrast to revenue-oriented forms of agriculture where object is to add more inputs with the hope increasing yields and profits. CER is the key measure for determining input-use efficiency.

One cited example of the usefulness of the CER lies in low-intensity plantations of shaded coffee. Based on Central American data (Perfecto et al., 1996), the costs for a coffee monoculture are $1740 and for the equivalent area shade system are $269. The resulting CER is

$$CER = \$1740/\$269 = 6.47$$

The unrefined CER value shows that, per area, the shade system is almost 6.5 times more efficient in term of added inputs.

This brings on the notion of two opposing economic options. One involves a greater level of inputs, increased revenue, and a greater profit. The other entails less inputs brought about by more reliance on natural dynamics and, through this, a greater profit.

As with the other ratio measures, the CER has broad and specific use variations. For the CER, adjusted values may be of greater interest that the raw or unaltered form. As a mirror to the land equivalent ratio, the CER has a landscape equivalent. The CER can be utilized for specific plantings, i.e., the cost savings from certain intercrops.

Adjusted CERs

Alone, the unrefined CER value is not very meaningful. This is resolved by adding the relative total value (RVT) to produce the RVT-adjusted CER. The resulting equation is

$$CER(RVT) = (c_a/c_{ab})\ RVT$$

For this, higher numbers are better, if the system is superior to a monoculture in input and land use efficiency both the CER and RVT will be greater than one, e.g.,

$$CER(RVT) = (1.5)(1.2) = 1.8$$

The notion is that high CER compensates for a low RVT, a low CER compensated for a high RVT. As an example, i.e.,

$$CER(RVT) = (0.75)(1.2) = 0.9$$

shows, on balance, a situation where the cost savings do not compensate for the productivity losses.

Using the coffee shade example presented above, the

$$CER(RVT) = (c_a/c_{ab})\ RVT = (\$1740/\$269)(0.22) = 1.42$$

Despite the lowering of expectations through a yield reduction and a low RVT (0.22), the cost gains are exceed the yield losses. This indicates favorable cost orientation.

In another version, LER can be substituted for the RVT to arrive at the LER-adjusted RVT. The purpose is to mitigate the effects of selling price on the analysis. This line of development continues as part of cost and economic orientation.

Economic Orientation

In evaluating agrotechnologies, it is important to determine if there are any essential resource efficiencies. These are shown by the LER. Likewise, cost efficiencies are measured by the cost equivalent ratio (CER). The overall economic gains, i.e., profitability, are revealed through a LER-CER comparison. There are other factors involved that relate to economics of design modification. Economic orientation measures if a system is revenue or cost oriented.

Revenue orientation puts the emphasis on increasing productivity and revenue. The hope is that costs will be less and this will result in a profitable outcome.

Photo 3.2. A coffee plantation where shade is indicative of cost orientation.

The opposite occurs with cost orientation. Higher yields are encouraged, to a point, but the emphasis is more on lowering costs. The hope that yields and subsequent revenue will still exceed the now lowered costs. As a result, a profit will be realized.

When calculating economic orientation, profit is the spread between the RVT and the CER. Reaching this point, it is better to switch to financial criteria and employ monetary units.

Importance

Economic orientation is more than an agroecosystem measure, it is an alternative direction. Take the case of Genetically Modified (GM) crops. These are more than advertized high productivity. They require large amounts of inputs to reach their promised yields. Without these, or with lesser amounts, traditional varieties may yield better. GM crops are also not bred to tolerate competition, either by way of weeds or in an intercrop.

Photo 3.3. A coffee planatation where less shade shows revenue orientation.

These plots become revenue-oriented. There are few other options other than revenue-oriented model.

Crops are not the only examples of revenue orientation. Hog/pig farming can be undertaken through intense or less intense methods. Revenue orientation occurs when the animals are fed, watered, and raised in total, often crowded, confinement. The costs are high, but so is the level of production and the revenue.

In contrast, free-range hogs, those that live outside, are still fed, but allowed to forage for supplemental food in pastures and/or forests. The production is lower as overstocking can lead to environmental damage. Adequate or significant profit is possible as per unit costs, both in equipment and feed, are reduced and, in some cases, eliminated, e.g., there is less need for barns and other buildings.

The same holds true with cattle raised in feed lots. The reduced-cost alternative is pasture-fed animals.

For crops, hogs, and cattle, the assumption is that greater revenue, i.e., revenue orientation, equates with greater profit. This does not always hold true.

Inter-Crop Comparison

Economic orientation is a comparison between a current practice and a proposed agrosystem modification or a biodiversity adjustment. In practice, the comparative base is narrow.

Economic orientation can become a universal measure. This is done by identifying a common base system, e.g., a maize mono-crop, and, through financial analysis, rank each crop and agrotechnology.

This is often accomplished informally. Users know that grazed natural pastures and forest-tree plantations are by far more cost oriented than paddy rice production. It would be nice, for landscape decision purposes, to have a formal numbers and these come through the wider use of economic orientation.

Calculated

Orientation, whether revenue or cost, is determined by comparing this against a monoculture of the primary species. The equation for finding the Economic Orientation Ratio (EOR) is

$$EOR = ((p_a Y_{ab} + p_b Y_{ba})/p_a Y_a) - (c_a/c_{ab})$$

For this, Y_{ab} is the yield of species a growth near species b, Y_{ba} is the yield of species b growth near species a, and Y_a is the comparative, size and site, monocultural yield of species a. The selling prices for theses yields are p_a and p_b for species a and b. The costs of production for species a are c_a, the joint costs, species a raised with species b, are c_{ab}. Since the

$$RVT = (p_a Y_{ab} + p_b Y_{ba})/p_a Y_a$$

and the

$$CER = c_a/c_{ab}$$

the economic orientation ratio can also be stated as

$$EOR = RVT - CER$$

Again, Utilizing Central American data on shaded coffee (Perfecto et al., 1996), the resulting EOR is

$$RVT = (p_a Y_{ab}/p_a Y_a) = \$314/\$1397 = 0.22$$

$$CER = c_a/c_{ab} = \$1740/\$269 = 6.46$$

and

$$EOR = RVT - CER = 0.22 - 6.46 = -6.24$$

The negative value of the EOR indicates cost orientation. The value, greater than six, shows very strong cost orientation. Although the analysis indicates that the coffee plants do not have a positive productive relationship with the shade trees, this drawback is more than compensated for in that cost inputs are being utilized with a high degree of efficiency. The above is a fairly conclusive example, demonstrating, among other things, why shaded coffee would be popular with poor farmers, those unable to afford expensive inputs.

In Use

To start, there is no indication that revenue-oriented agrosystems are any more profitable than cost orientation. A lot depends on the selling price of the crop or crops and the cost of the inputs. With this context, there are considerations and pros and cons regarding each system.

Revenue Orientation

Revenue-oriented farming can apply to a few plots on an economically-balanced farm or this strategy can encompass a wide area with only one crop and one agrosystem type. There is no problem with a few revenue-oriented plots in an otherwise balanced landscape. In this case, there might be spillover from the ecologically-rich areas to the ecologically-lacking, revenue-oriented plots. These mixed farms have room for, and can even welcome, local flora and fauna.

Issues arise when revenue-orientation reach the scale and magnitude where nature and natural dynamics are totally excluded, replaced by chemical inputs. By extension, natural flora and fauna are in-the-way additions. This is loosely defined as being unsustainable. This is also commonly referred to as the factory farm.

With the factory farm, profit and return on investment constitute the sole objectives. With the financial objectives, along with the missing natural dynamics, environmental concerns are often subjugated.

There are drawbacks with a one crop farm and a totally reliance on revenue orientation. Much of this relates to risk. High input systems quickly become unprofitable if crop selling prices drop. There are financial dangers when input costs rise or when yields are low or non-existent. This can happen with adverse weather, disease, and/or insect infestation. These can reduce crops yields while not diminishing associated costs. In these cases, the greater initial investment translates into a greater financial loss.

As farm field grow is size, the ability to closely monitor diminishes. Classic Integrated Pest Management (IPM) is harder to implement and, when problems are detected, larger areas are sprayed. This adds to the per area, and per output, costs.

There are some cost gains. Specialized farm machinery can be purchased and better utilized across a greater cropping area. Much the same applies to management. With a single crop supplied with outside inputs, management can operate from a far smaller knowledge base. This allows less skilled employees, using operating manuals rather than experienced-based judgment. This even allows management from afar.

For governments and the economy as a whole, there are macroeconomic gains with revenue orientation. Commerce is at a higher level, specifically in the selling of larger farm machinery, agrochemicals, and other inputs as well as from the transport and processing of outputs. With these cash flows, there are increased opportunities for taxation.

These are reasons that, in some countries, governments provide selling price floors, crop failure insurance, and farm subsidies. This takes some of the risk out of revenue orientation and fosters revenue-oriented agriculture.

For some, these gains resonate. Large farms have become revenue-oriented, one-crop monocultures, e.g., the maize and soybean producing regions of North America and the soybean regions of Brazil. These are also found in Europe, Australia, and other parts of the world.

Cost Orientation

With far fewer opportunities for ecological expression, revenue orientation this is far from the agroecological ideal. As the opposite of revenue orientation, cost orientation is also financially viable and can encompass large or small areas.

The large treecrop plantations found in tropical regions, e.g., rubber and oil palm, and possibly some of the large forest-tree plantations follow this model. Because of the inherent understory biodiversity, these have more of an agroecological imprint than the large-area, weed-free, pure monocultures.

Within the larger farm landscape, it is best to have a mix of plots, some revenue-oriented, some cost-oriented. The balance is needed as landusers may not have the ability to purchase inputs, including labor, for an entire farm. As not all plots are planted or harvested simultaneously, mixed orientation spreads labor requirements. This can mean fewer people are employed, but for a longer time period.

Cost orientation embraces various inter-plot synergies. These are within the farm landscape and include the ecology associated with inter-plot rotations, inter-plot, ecological spillover (e.g., insect control), and/or from utilizing manure from on-farm livestock.

Following economic protocol, the strategy is often to put ample resources on the most fertile, well watered plots and to make the other plots cost oriented. The result, if crop selling prices are high and the cost of inputs low, then the revenue-oriented plots become the profit centers. If not, the cost-oriented sections of the farm may provide the greater profit.

The overall advantage are well documented. With their low inputs, cost orientation can remain profitable even when crops sell for less money. Also, since inputs are low and/or not externally purchased, inputs prices have proportionally less or no influence on the financial aftermath. Also, if the weather is not overly favorable, the financial loss is not as high as when much more money had been invested.

Cost orientation is not only associated with small farms. This can occur with large farms. An example is the cocoa-growing region of Brazil. The cocoa is tree shaded where, if revenue orientation was sought, these would be in full sunlight sans trees.

There are other motives. These are the financial, practical, political, climatic, topographical, or even aesthetic (a more picturesque setting) considerations that encourage mixed orientation. This text stresses the revenue and cost gains from an agroecological-balanced landscape. The worldwide norm seems to be for farms to be a mix of cost and revenue oriented agroecosystems.

Possibility Curves

The above ratios are diagnostic, these serve to economic position and understand why an agroecosystem finds or will find favor. In economics, the next step is to optimize the system with regard to some or all of the variables. There are a series of curves that do this. These start on the revenue side.

The Production Possibilities Curve

For any intercrop, there arises the question of the optimal planting densities. This question is usually answered by way of the Production Possibilities Curve (PPC). This curve can be quantified with yield data or expressed in ratios.

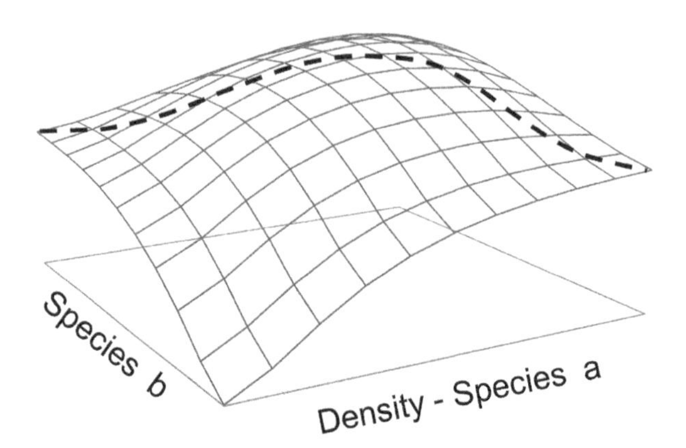

Figure 3.1. 2-D and 3-D production possibilities curves where the 2-D would run along the ridge line of the 3-D surface. Figure 3.2 shows this in overview.

The simplest form, this is two dimensional. This has, as axises, the LER values. This classic 2-D form is shown in Figure 3.1. The optimal planting densities are determined with price ratios. These determine the point on the curve which is best for a current price ratio (as in Figure 3.1).

Although useful, only a few have been derived, e.g., Ranganathan et al. (1991). One reason is the amount of data required. Even a not-so-statistically-significant curve can require an experimentally large, difficult-to-obtain data set.

In the complete form, the PPC is a three-dimensional function. For a two-species intercrop, the x- and y-axes are planting densities. The upper, z-axis is output measured as an LER value.

Derivation of the three dimension form requires correspondingly larger volume of data. Offering more information, the three-dimensional version is more informative than the two-dimension form. Without applied examples from which to derive a 3-D surface, the information is unavailable.

The PPC can be based on actual yield numbers or, as done in this text, use ratios with a zero to one range. For a ratio-based revenue version of the PPC, the curve can be based on the crop selling prices where the RVT is the z-axis.

The Cost Possibilities Curve

As many seek cost-oriented systems, there exists the Cost Possibilities Curve (CPC). Lacking the needed field data, this remains an abstract construct. Still, cost savings occur with intercropping, e.g., where one weeding benefits both crops. From what is known, a curve can be derived. This would have, for a two-species intercrop, planting densities as the lower, x- and y-axes, total costs would distinguish the z-axis.

The Profit Curve

Revenue and cost curves are only a subtraction away from an estimate of profit. Both the revenue and cost curves can be expressed points on the three-dimensional curves and these translate into number matrices. Revenue minus costs equal profit and this occurs with revenue and the cost matrices. The high point in the derived profit curve. Through ratio lines, the optimal planting densities are determined.

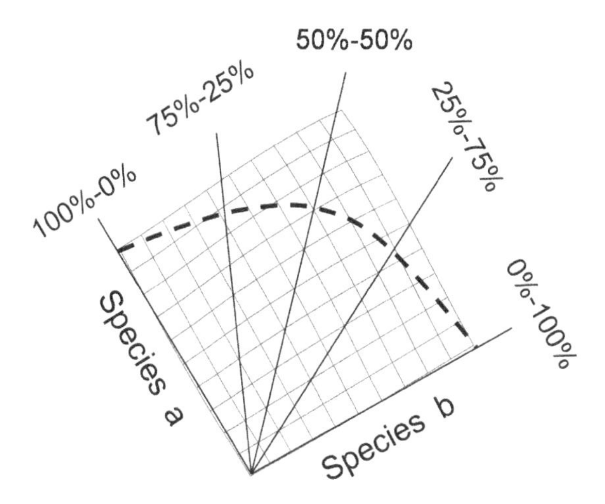

Figure 3.2. Ratio lines illustrated. This have been added to the 2-D curve and the 3-D surface in Figure 3.1.

Ratio Lines

On the 2-D and 3-D PPC curves, ratio lines radiate out from the origin (0,0 point, lower left) of the production, cost, and profit curve. The axises of the different curves represent different mono-planting densities. The line that is midway between the axises is the 100%-100% or 50%-50% ratio line. The different values are shown below.

for the primary species plants per area (and the percent)	and for the second species plants per area (percent)
10,000 (100%) <-------------------------	-->20,000 (100%)
7,500 (75%)	15,000 (75%)
5,000 (50%)	10,000 (50%)

2,500 (25%) 5,000 (25%)

2,000 (20%) 4,000 (20%)

1,000 (10%) 2,000 (10%)

The 100% values, as the optimal monocultural planting densities, are at the end of, and anchor, these values. Similarly, other ratio lines are possible. Shown below and in Figure 3.2 are sample points in the 100%-50%, 100%-25%, and 25%-100% ratio lines.

100%-33% = (10,000 - 3,333) or (5,000 - 1,666) or (2,000 - 666)

100%-25% = (10,000 - 5,000) or (5,000 - 2,500) or (1,000 - 500)

25%-100% = (2,500 - 20,000) or (10,000 - 1,200) or (5,000 - 600)

For optimal profit, the line which crosses through the highest point in the three-dimensional profit curve represents the best planting ratio and densities.

Output/Input Ratio

Almost all the ratios go to strengthening micro-economic analysis. The exception is the output/input ratio. This is often more of a macro-economic measure. However, it does have value in gauging the amount of yield-producing ecology on a farm or within classes or categories of farms.

The output/input ratio looks at how efficient are the yields in terms of outside resources expended. Many see a world where the food supply depends on the flow of mined minerals, specifically phosphorus and potassium. If these commodities, as well as the array of agrochemicals, are stopped, how would agriculture fare?

In many aspects, this addresses the larger notion of food sovereignty. Food may be the most important human need and the right to food should not be subject to any strife this disrupts the supply or flow. Peoples and regions should be food self-sufficient.

The output/input ratio is the total value of the outputs divided by the inputs used. This is not a financial evaluation. Labor is not included, tractor purchase cost amortization and fuel would be an input. Inter-plot and intra-farm transfers of nutrients, e.g., mulch and manures, are not part of the calculation. Also, the application of agricultural waste, if locally sourced, would not be an input. Fertilizers and other agrochemicals are included.

Out of necessity, this requires monetary units. The output/input ratio can be a crop, plot, and/or a farm-wide measure.

This is a form of risk, i.e., becoming to depending on and losing the means for food production. Agronomic self-sufficiency starts when imports are used judiciously or, per unit of output, utilized very efficiently.

This ratio quantifies regional or country-wide food security and was formulated to measure food sovereignty. One strength of agroecology is a reduced reliance on ex-farm imports. Farms with lower output/input ratios may use inputs that are locally or on-farm sourced and/or may exhibit high degree of plot or landscape-based agroecology.

Summary

This chapter has looked at ratios without much reference to the financial aspects. This is because ratios are a simple diagnostic tool that provide a range of insights that are somewhat obscured when currency units are affixed. The transition from ratio analysis to financial analysis is easily accomplished.

Landscape analysis is major topic not touched on here. This is introduced later in this text (Chapter 12). Risk also merits, and constitutes, advanced analysis (Chapter 10).

4

Plot Agroecology

Chapter Preview

This chapter begins the process of deriving universal equations. The starting point is the land equivalent ratio and the allocation of essential resources between competing plant species. There are two sides to this, (1) competitive partitioning and (2) facilitation where competitive partitioning subdivides into (a) separate sources and (b) functional non-linearity. These are reviewed and hypotheses proposed on the relationship between these two influences.

Chapter Contents

Accumulation-Concentration
Site Improvement
Parasitic
Non-Resource Facilitation
Economic Facilitation
Net Effects
Outcomes

Introduction

As outlined in Chapter 2, the agrotechnologies are a means to grow food, fiber, and fuel. As stated, these can be rigid systems where the internal designs are preset, i.e., the species, spacings, and timing, and, by extension the internal agroecology. Also preset is the expected economic outcome. At some point, either during a research stage or through on-farm manipulation, certain design variables and, by extension, the economic results, will have been decided. This has not happened with all the agrotechnologies.

Monocultures are uncomplicated and comparatively well studied. In comparison, the implementation directives for bio-complex agrotechnologies can be quite rudimentary. For a few, only scant details guide use. These remain in flux.

The initial focus will be on the plot-internal agroecology. These are the mechanisms that govern how interacting plant species relate to each other and the site resources. Specifically, these are the mechanisms of competitive partitioning and facilitation.

Depending on the agrotechnology, an agroecosystem can be a mix of annual crops, mixed fruit trees in an orchard setting, or a combination of perennial wood-producing species as in a forest plantation. These can also be some combination of orchard species, forest trees, and crops. All share the same cross-species dynamics. The simplest expression, and the starting point for this discussion, is the two-species intercrop.

Two ecological mechanisms, competitive partitioning and facilitation, are discussed in this chapter. A third, that of exclusion, is reserved for Chapters 7 and 8. In agroecology, the importance of exclusion lies in weed control.

Essential Plant Resources

In agroecosystems, there is competition for light, water and nutrients. This is expressed as

$$Y = [f(L), f(W), f(N)]$$

where Y equals crop or plant yields and yields are some function of light, *f*(L); water, *f*(W); and nutrients, *f*(N). This is a general equation where, if expanded, nutrients break down into nitrogen, phosphorus, potassium (N, P, and K) and a host of trace elements. The trace elements include boron, calcium, copper, iron, magnesium, manganese, molybdenum, sulfur, and zinc. For completeness, soil properties might be part of an expanded expression, e.g., bulk density and organic content.

Light is another critical input. Where inter-species apportionment is required, light can be subdivided into vertical (midday) and horizontal (early morning and late afternoon). Not mentioned in the above equation is CO_2. Since this usually does not play a large part in the dynamics of an intercrop, this is considered an unstated constant.

Inter-Species Interactions

When different species compete, two core interactions occur. These produce a significant outcome. The first is competitive partitioning, the second, facilitation.

Competitive partitioning happens when two or more individual plant species, grown in close contact, compete for the same essential resources. The advantages are least observable when the two interacting species are genetically and characteristically similar, most exploitable when the competing plants are very different in their physical presence and/or their essential resource needs.

Facilitation is where the presence of one plant species benefits another. This follows the ecological definition where '*... species interactions benefit at least one of the participants and cause harm to neither*' (Wikipedia, Ecological Facilitation, 2013). In agroecology, the participants are the plants. When dealing with essential resource, facilitation adds to the amount of available water and/or increases the level of essential nutrients, this can result in increased yields.

There are other forms of facilitation. This includes guarding productive plants from climate-related threats and protecting these same species against unwanted flora and fauna, e.g., weeds, herbivore insects, grain-eating birds, etc.

For all these categories, there are caveats and exceptions. In the majority of cases, these categories hold and are an analytical foundation.

Underlying Analysis

Competitive partitioning and facilitation are part of a system of equations. In biculture form, the LER is

$$Y_{LER} = (Y_{ab}/Y_a) + (Y_{ba}/Y_b)$$

where

$$(Y_{ab}/Y_a) = 1 - P_{ab} + F_{ab}$$

and

$$(Y_{ba}/Y_b) = 1 - P_{ba} + F_{ba}$$

For this, P_{ab} represents, through competitive partitioning, the allocation of essential resources to species a. P_{ba} is the share of these resources going to species b. F_{ab} are the facilitative effects species b has on species a, F_{ba} are the facilitative effects of species a on species b.

As a determinate of the LER, the values represent the site resources. The values are, at this stage, single points. This aspect is developed further in subsequent chapters.

Combined, these are

$$Y_{LER} = (1 - P_{ab} + F_{ab}) + (1 - P_{ba} + F_{ba})$$

A value of one (1) represents the maximum LER without competitive partitioning and facilitation. These changes transform the LER equation from an economic to an agronomic expression.

This function combines with the essential resources equation as

$$Y = [f(L), f(W), f(N)]$$

Expanded, this is

$$f(L) = f_{Light}((1 - P_{ab} + F_{ab}) + (1 - P_{ba} + F_{ba}))$$

$$f(W) = f_{Water}((1 - P_{ab} + F_{ab}) + (1 - P_{ba} + F_{ba}))$$

$$f(N) = f_{Nutrients}((1 - P_{ab} + F_{ab}) + (1 - P_{ba} + F_{ba}))$$

Normally, yields and LER are minimum of one essential resource. This is where one resource constrains yields. This is termed the limiting resource. With this, the equation is

$$Y = \min [f(L), f(W), f(N)]$$

Bio-Efficiency

There are increases in the resource capture that are solely the result of species diversity and not attributable to any one species. These involve the acquisition of light and/or water.

In theory, having more ground cover and more root channels for infiltration should result in more rainfall being retained. Upon examination, this does not always hold true. When studied, water capture by intercrops does not differ much or is slightly better than that of monocultures (Willey, 1979; Morris and Garrity, 1993b).

This same non-attributable, general effect should augment the interception of light, i.e., with more biomass, less direct light would strike bareground. Again, this is fairly minimal. Although efficiency gains are real, they are generally minor. More often, the effect is considered implicit in competitive partitioning or facilitation.

Competitive Partitioning

In the competition for light, water and nutrients, there are a range of dynamics at work. These start with competitive partitioning (P_{ab} and P_{ba} above). This is also termed competitive production.

For competitive partitioning, there are a number of contributing factors. The most significant gains result from the non-linearity of the production functions. If one of the component species has separate sources for one or more essential resources, this adds to the non-linearity gains.

Functional Non-Linearity

Competitive partitioning is, in large part, defined by functional non-linearity. When two plant species interact, they must divide the acquirable essential resources. With linear production functions, no gains occur. Non-linear production functions results in LER gains. This mechanism has not been studied, but it can be demonstrated.

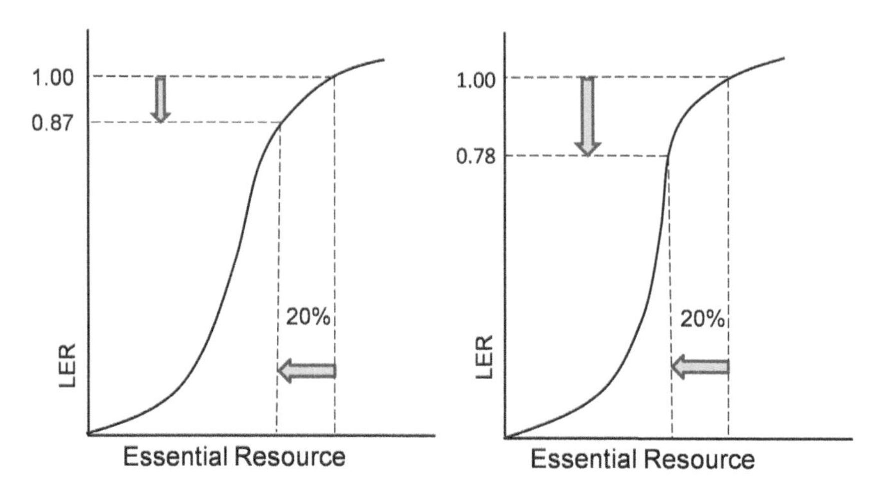

Figure 4.1. Functional non-linearity is demonstrated by way of hypothetical production function for two species (left and right) and one essential resource. This shows where the addition of a second species takes 20% of the resource from each plant. This puts the LER at a more-than-acceptable 1.65 (0.87 + 0.78).

When a field is planted, the subsequent harvest does not drain the area of all resources. A crop takes a roughly estimated 20% of the resident minerals (Palm, 1995).

Using this as an illustrative benchmark, Figure 4.1 shows hypothetical productive functions for two species (left and right) and one essential resource. The monocultural yields are labeled 1.00 on what would be a very resource rich site.

When inter-planted, the yields are diminished when each species takes about 20% of one or more resources available to the other species. The assumption is that each species is planted in what would be the optimal monocultural spacing. In Figure 4.1, this results in an LER value of 1.65 (0.87 + 0.78). This would be a very good outcome.

This example can be carried further. In Figure 4.2, the site is less resource affluent and the base monocultural yields are therefore less (still labeled 100%). Given the steep slope of production function at this point, a 20% reduction in a limiting resource dramatically drops yields. The LER is an unacceptable 0.55.

This drop off in the LER does not eliminate the biculture. The simplest way to boost the LER is to reduce the planting densities for each component species. Better design, through additional mechanisms and inter-species complementarity, is another option.

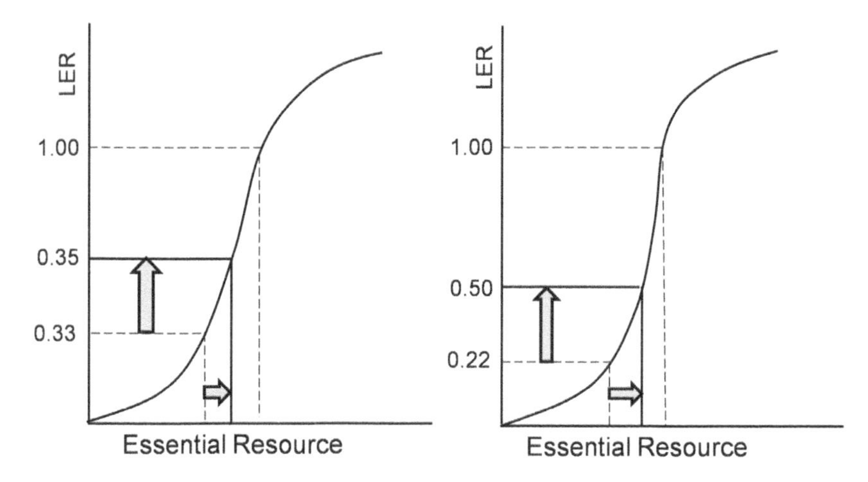

Figure 4.2. With a more resource limiting site, a 20% mutual reduction due to resource competition plunges yield from a mono-crop value of 1.00 to a bicultural value of 0.55 (0.22 + 0.33). With the addition of separate sources. The LER increases to 0.85 (0.35 + 0.55). This contrasts to the high values for Figure 4.1.

Separate Sources

Often plants to do not directly compete for one or more essential resources, but might find a separate source or origin. As a component of competitive partitioning, the relationship is

competitive partitioning = – functional non-linearity + separate sources.

Within the upper, less sloping reaches of the production function, functional non-linearity is generally greater than separate sources. In the lower, steeper range, functional non-linearity becomes negligible, but separate sources remain. One very poor sites, where there are few site resources, separate sources zeros out. The same holds true when sites are very rich and resources abundant. In a biculture, separate sources, sans facilitation, cannot elevate the LER above 2.0.

The dynamics involved, itemized, are where

(1) one or more essential resources are utilized in a different time or season,
(2) the resources are utilized at the same time or season, but are collected in different zones, e.g., plant root profiles strata (for nutrients or water) or canopies (horizontal or vertical), and/or
  (a) nutrients are in different soil strata or plants collect different forms of light
  (b) resources are obtained from different origins (e.g., in-soil nitrogen or through nitrogen fixation).

Timing

The classic example of timing is the pairing of tomato with lettuce. The lettuce grows fast, is harvested first, and uses light, water, and nutrients early in the growing season. Tomatoes are slower to mature and require more light, water, and nutrients later in the season. These pair well primarily due to the timing of their uptakes.

This can be intensified by planting the fast mature species early, wait some days or weeks, then plant the other. This increases yields, but can also add to the cost.

The same mechanism occurs with early and late maturing perennials. This is where a species that grows fast early in the season when paired with a species that experiences greater growth latter in the season. This can occur when one tree species puts on foliage early in the spring, while another waits until late spring or early summer.

A similar dynamic occurs with hedgerow alley cropping. The hedge produces branches and leaves during dry season when no crop is grown. Before the wet season begins, the leaves are cut to provide biomass and nutrients for the rainy-season crop.

Photo 4.1. A maize with cassava intercrop that exhibits a temporal dimension. In this case, the maize has been harvested giving the inter-planted cassava free access to the remaining site resources.

Different Derivations

Where light is the limiting resource, one competing species might seek vertical light, another might want horizontal light. A compelling example is an onion/cabbage intercrop. Onions, with their vertical canopy, are a seeker of horizontal light. Cabbage, with horizontal leaves, captures vertical, midday light.

For nutrients, some plant species may look more in the deep soil strata, other species may want surface nutrients. The same may hold true with water. Annual plants with mainly surface roots may subsist on frequent periods of light rain. These crops may be paired with deep-rooted perennial trees. These trees would gain from access to deep ground water. This situation, as a parkland agrotechnology, is found with trees and crops (millet and sorghum) in sub-Sahara Africa.

Different Needs

Different needs can occur when one species seeks a more of one resource, less of another. If one intercrop component seek more nitrogen, less potassium, species complementarity occurs when a co-planted species wants more potassium and less nitrogen. All else held equal, this should up the LER.

This also applies to light. Many light-demanding species have open canopies where much light reaches the ground. A shade-resistant plant species, one with low light needs, can serve as an understory plant. This is the basis for heavy shade systems.

Agroecological Niches

In agroecology, separate sources is the best approximation of a niche. In ecology, the niche is an abstract concept that '*... describes how an organism or population responds to the distribution of resources and competition*' (Wikipedia, Ecological Niche, 2013).

In agroecology, the niche is less abstract and, given enough data, it can be quantified. In Figure 4.1, it was assumed that there are no niche gains associated with functional non-linearity. Figure 4.2 changes this assumption. Adding separate sources to functional non-linearity can hike the LER.

In the best case, each component species would draw all its key resources from sources to which the second species does not have access. When this happens, there would be no yield reductions and, for two interplanted species, an LER value of 2.0. This represents niche gains of

100%. With two inter-planted species, LER values above 2.0 come about only through facilitation.

There might well be a rare exception. This is where two species are present but, due to varietal difference within one of these, this system functions, i.e., gathers resources, as three species might.

With a three species polyculture, the potential LER would not reach 3.0. This is because of niche crowding. With a four-plus polyculture, sans facilitation, the 3.0 value might approximate as the upper limit. In most situations, a lesser value would likely prevail.

Facilitation

More of the gains through competitive partitioning come when soils are comparatively rich and well watered. Where one or more nutrients or, in some cases, water are in short supply, plant-on-plant facilitation becomes the main mechanism for achieving acceptable LER values. In the expanded LER equation, these are the values F_{ab} and F_{ba}.

In their mechanism and functional range, these are quite different. Competitive partitioning relies on functional non-linearity to achieve high LER values. Separate sources adds to this. If the functional non-linearity results in a low, and undesirable, LER, i.e., LER < 1, separate sources can lift this number (as in Figure 4.2).

This is not the case with facilitation. Under this mechanism, the facilitating species most derive one or more resources from separate sources in order to supply the receiving species with one resource. Simply stated, without separate sources, resource facilitation would not exist.

There are a number of facilitatory mechanisms. By convention, resources facilitation is always positive. This does not mean that the presence of a potentially facilitative species always results in a yield-positive outcome.

A nitrogen-fixing plant may make nitrogen available to an abutting species. There might also be an accompanying negative, non-facilitative effect. The nitrogen-fixing species may also usurp water. Although the conditions exist for nitrogen facilitation, competition for water may cause a net yield loss with the primary species. Facilitation might come about with more rainfall.

The most common form of facilitation is where the primary species receives help from a secondary species. This is one-way or, in ecology, commensalistic facilitation.

Also encountered is mutualism. This is where there are cross effects, a secondary species aids the growth of the primary species (through one resource) and, at the same time, the presence of the primary species assists the secondary species (by way of a second resource).

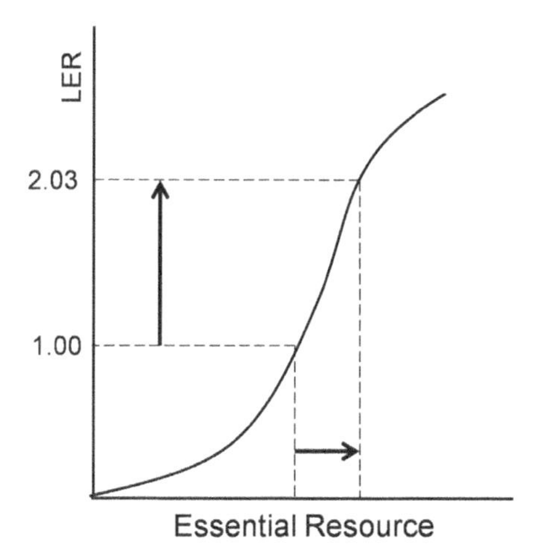

Figure 4.3. The effects of facilitation where a comparatively small increase in the limiting resource to the primary species can dramatically increases the yield. Showing only the yield gains from the recipient species, the LER goes from 1.00 to 2.03.

Yield Gains

When sites are resource poor, there are ample opportunities for facilitative gains. Figure 4.3 shows the resource poor range and large jumps in single-plant productivity that can result from adding a comparatively small amount of one essential resource. This is borne out in practice. The highest reported LER increases, about 3.5, are obtained when in-plot trees provide nitrogen to a nitrogen-demanding crop (Ong, 1994; de Moura et al., 2010).

Facilitation also applies to moisture dynamics. Windbreaks on a Mongolian plain decreased wind by 32 to 38 percent, brought about increase in soil moisture by 3 to 6 percent, decreased in summer temperatures by 0.1 to 0.7°C, and an increased in winter temperatures by 0.5 to 1.6°C. Seemingly small, this increased maize yields by 64% and millet yields by 70% (Zhang Fend, 1996).

Growth and Yield

Resource facilitation occurs by supplying one or more essential resource to the primary species. The norm is one. More than one might occur when green biomass is cut and applied to a companion crop. Ideally, this is includes the yield-constraining resource for the primary species.

Figure 4.4 shows when the addition of one resources results in the greatest yield gains. There a various ways to achieve this. The mechanisms that can produce yield gains are

- nutrient capture-transfer
  - nutrient pump
  - airborne nutrients
  - fixation of nitrogen
  - chemical conversion
  - timing of release
- water capture-transfer
- retention
- accumulation-concentration
- site improvement
- parasitic

Nutrient Capture-Transfer

On many sites, productivity is constrained by mineral resources. The resource or resources that constrain plant yields can be nitrogen, phosphorus, potassium, or some trace element. There are a variety ways that one or more facilitative plants can supply nutrients to a primary species.

One mechanism is the nutrient pump. Basically, a deep-rooted species is paired with a shallower-rooted primary species. The hope is that deep rooted species will find, in the deep soil strata, essential elements and these are brought to the surface and incorporated into the leaves and stems of facilitative plants. Subsequently, these are released through leaf fall and decay. Although the nutrient pump is confirmed (Velasco et al., 1999), the effect on productivity is not known. The best option for implementation is to add deep rootedness as a desirable characteristic for a companion species.

Another mechanism of facilitation is the capture of airborne nutrients. This is significant. A fundamental example is where nitrogen-fixing plant species find this element in the air (N_2 for legumes vs. NO_2 for non-legumes).

Airborne nutrients, other than nitrogen, can be captured from the air. Impacted dust which is slowed and allowed to settle through the wind-blocking trees. This can add a range of elemental nutrients. There are the bromiliades, plants that survive solely through the aerial capture of water and nutrients. These plants, through death, fall, and decay release minerals under the trees in which they live.

Nutrients can be bound in plant-inaccessible chemical compounds and/or physically trapped in rocks. Facilitative plants, those that breakdown these associations, can add more available nutrients to the primary species.

One lesser known facilitative mechanism is the timing of release. Nutrients are often trapped and are inaccessible in decaying plant remains. If the decay process is accelerated, these could provide more immediate yield benefits. This is often accomplished in a more shaded, more humid agroecosystem.

Water Capture-Transfer

Essential nutrients are only part of growth facilitation. A close cousin of the nutrient pump is the hydraulic lift. Again, a deep-rooted species pulls water from a lower soil strata. This benefits the primary species through release into the soil and subsequent capture or by greater humidity and less aboveground drying. As with the nutrient pump, the effects on yields are not known and most likely small.

Retention

Another aspect of facilitation does not involve finding additional water or mineral resources, but keeping that available from being lost.

One aspect of retention is in countering erosion. A ground cover or series of barriers do this. Facilitation come by way of a covercrop species that uniformly blankets the soil. It also comes through erosion-blocking barrier hedges.

Not all retention involves holding soils. Plants, either closely spaced or as wind shelters or windbreaks on nearby plots, can increase the humidity of a plot. This retards drying, thereby reducing the moisture requirement. Although the effects of micro-climatic change are small, these have been shown to be a significant influence, e.g., Zhang Fend (1996).

Accumulation-Concentration

The classic fallow is based on the notion that, over a suitable time period, nutrients buildup in the soil. Natural accumulation is accelerated by increased biodiversity and/or by including fast-growing, high-biomass species. This is a form of facilitation.

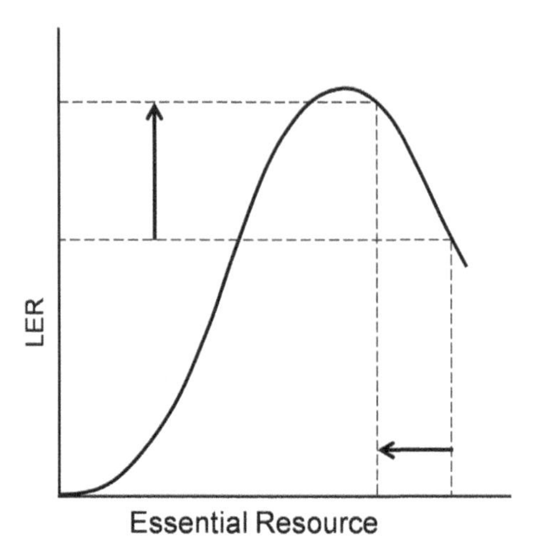

Figure 4.4. A second, less encountered, version of facilitation where the removal of a resource increases the yield.

Site Improvement

At times, sites can be unsuitable because of a poor site. Soil conditions, such as bulk density, porosity, pH, etc., are changeable by way of facilitative species. Improvement can be a long-term project.

Another improvement, sometimes called subtractive facilitation, is the removal of an essential resource. Although examples are less numerous, the affect on output can be equally pronounced. This is when site resources are absorbed by the facilitative species and, as a result, this results in a yield gain. This only occurs when one resource is in situational oversupply and this has a negative impact on growth and yield of the primary species.

This second facilitative mechanism is shown in Figure 4.4. The overabundance of one resource, as shown in the downward sloping curve, is adversely impacting potential yield (right portion of the curve). Taken away, yields maximize. There are prominent examples, maize and a few other crops are very susceptible to water-saturated soils. When this occurs, less moisture produces greater yields.

Parasitic

A few plants directly feed off, and obtain nutrients from, other plants. These are the parasitic plants. A very minor category in agroecology, there are a few of commercial interest. The mistletoe is parasitic on the branches of oak trees and is sold as a decorative. An example with considerable monetary worth is the truffle. This is a belowground addition to tree-based agroecosystems.

Non-Resource Facilitation

There are forms of non-resource facilitation. Physical support is one type. This is where shrubs or trees provide support for vine crops, e.g., grapes, pepper, and kiwi fruit. Having the primary vine crop over top shrubs, dwarf or pruned trees is one option. A second option has a shade tolerant vine grown on trunks of tall, widely spaced trees.

Support can also be belowground. Orchard species can topple with a combination of wet soils, heavy fruit yields, and/or wind. An adjacent plant, usually a woody shrub, can, through interlocked roots, eliminate lodging.

Photo 4.2. Squash under maize. This is less a two-output intercrop and more an example of economic facilitation, i.e., where the squash suppresses weeds.

Economic Facilitation

Economic facilitation comes about in situations where there is no transfer of essential resources, but a second species provides economic gains. This is where there may be a net loss in yields and revenue, but this is more than made up by a cost reduction and a gain in profitability. Commonly this occurs when the facilitating species suppresses weeds. For example, when maize planted with red clover reduced maize yields slightly, but decreased weeds by 72% (Altieri, 1995).

Along these same lines, there are also facilitative gains when insects, plant diseases, winds, damaging animals, etc. are deterred. As variations of non-resource facilitation, these have bearing. Discussion is delayed to later chapters.

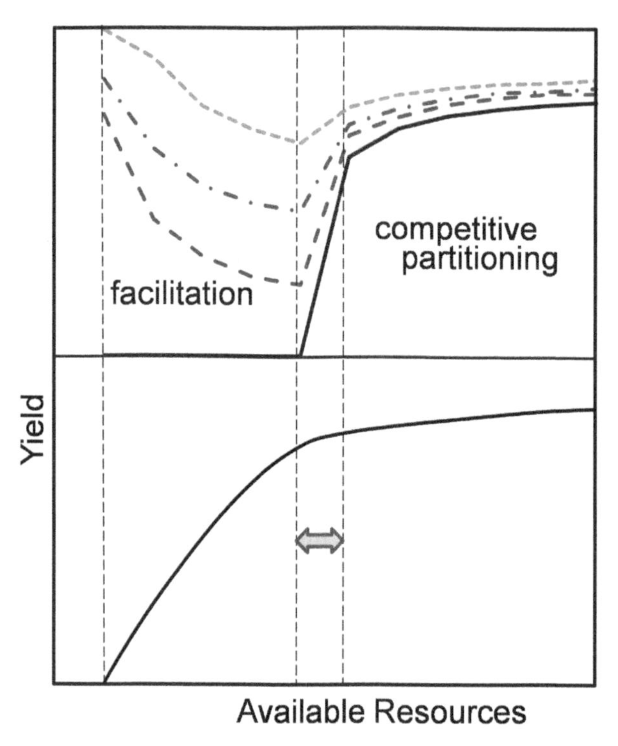

Figure 4.5. The theoretical relationship between competitive production (in this case, functional non-linearity) and facilitation. Comparing this to a single production function in what would be a bicultural planting (lower section), the potential strength of each effect is shown (upper section). The middle section (arrow) is where facilitation and functional non-linearity overlap.

Net Effects

This chapter has looked at competitive partitioning and facilitation. Competitive partitioning, through separate sources, can come about on rich and/or poor sites. Competitive partitioning, by way of functional non-linearity, occurs only on resource-rich sites. If a site lacks in one or more resources, functional non-linearity become less of an LER contributor, niche (separate source) gains become ascendent.

Functional non-linearity is strongest when, for the one limiting resource, there is little opportunity for separate sources. Functional non-linearity would not occur when each species has different source for the resource in question.

For sites lacking in one essential resource, the facilitative gains, if well directed, can be significant. Under resource facilitation, the supplying species must have a separate resource sources in order to transfer the one resource to the receiving species.

Although competitive partitioning and facilitation are not totally exclusive, there is some, and possibly an insignificant, overlap as these occur in different segments of the essential resource range.

In modeling these effects with varying assumptions, Figure 4.5 shows competitive partitioning, with most of the gains from functional non-linearity (upper right). Facilitation (upper left) is shown in varying degrees (low, medium, high). The sample production function (lower section) shows how competitive partitioning and facilitation relate to a production function and to the available essential resources.

Figure 4.5 assumes only one limiting essential resource (often nitrogen). Some sites are multidimensional in that there is a shift from one limiting resource to another and possibly on to yet another. There can also be a seasonal back and forth. This can muddy the situation. It can be stated that facilitative gains always trump, i.e., exceed, those from competitive partitioning.

Outcomes

The above expands the understanding and provides a foundation for LER-based intercropping research. Without such, plant/plant relationships can be misdiagnosed. Success of the classic maize/bean intercrop (with LER values that approximate 1.50) is often attributed totally to facilitation.

The seemingly obvious conclusion is that the bean crop is nitrogen fixing and this benefits nitrogen-demanding maize. In support, cases where nitrogen fixation potentially benefits an accompanying crop are not uncommon. Some of these combinations, those that feature highly positive LERs (in parenthesis), are

barley and fava bean (1.85)
cassava, groundnut, and maize (2.51)

cocoyam, groundnut, and maize (2.08)
groundnut and millet (1.26)
maize and pigeonpea (1.67)
okra and cowpea (2.62)

The counter argument is that high LERs might also be the result of competitive partitioning with or without separate sources (i.e., based more on functional non-linearity). In support, similar LER values are obtained with co-plantings sans a leguminous plant. Some published LER values are

maize and coriander (1.42)
maize and okra (1.75)
maize and potato (1.34)
amaranth and okra (1.91)

The lesson here is that there are dangers in relying too much on generalizations. This is further illustrated in the different outcomes that occur with different varieties of the same crop. For a maize/potato intercrop, one potato variety gave an LER of 1.34, another 1.06 (Kimani, 1987).

Despite the caveats, it is likely true that maize is a very complementary partner species. When intercropped, values of around 1.6 are feasible with a reported high of 2.6 (Seran and Brintha, 2010). Being a popular agronomic species which intercrops well, maize is comparatively well researched. Staple crops, such as cassava and sorghum, also seem to intercrop well, but are less studied.

For varying and obvious reasons (e.g., harvesting issues), some staple crops, e.g., rice, wheat, potato, seem more challenging as partner species. This has meant fewer intercropping possibilities and less bi-cropping research.

It is quite possible that the best outcomes occur when two species have a facilitative relationship when a key resource is in short supply. When this resource is relatively abundant, the two species exhibit positive non-linearity. For the other essential resources, there is less need and/or the plants have separate sources. Although not fully explained, this could be the reason for the popularity of the maize-bean intercrop.

Intercropping does not lack for species pairings (as demonstrated in Table 6.1). Underlying this, there will be patterns, maybe even guidelines, as to which species intercrop best. Until there is sufficient empirical data, the starting point is agroecological theory.

As seen, this is a complex topic where competitive partitioning and facilitation are, at the field level, far from explained. As is often the case, better conceptional understanding leads to better analysis. More development along this road, especially in explaining the minutia and nuances, will help in formulating more informative research and in designing more purposeful agrosystems.

5

Production Functions

Chapter Preview

This chapter looks at the production function in regard to competitive partitioning and facilitation. It is not a smooth road from single plant, single-resources function to the polycultural, multi-resource form. There are obstacles along this road. Despite this, a general, quantified equation is possible.

Chapter Contents

Introduction

The equations presented in proceeding chapters are mostly abstract. In going from the conceptual to the qualitative, the production functions must assume a more precise form. Non-linearity is a key aspect as this underlies competitive partitioning, facilitation, and the other gains from intercropping.

For the corresponding productions functions, there is little agreement as to which equation form or forms are best. In addition, there are differing opinions on the general shapes these functions might assume. Despite the many unknowns, there are clues that allow realistic production functions to be tentatively derived and parametrized.

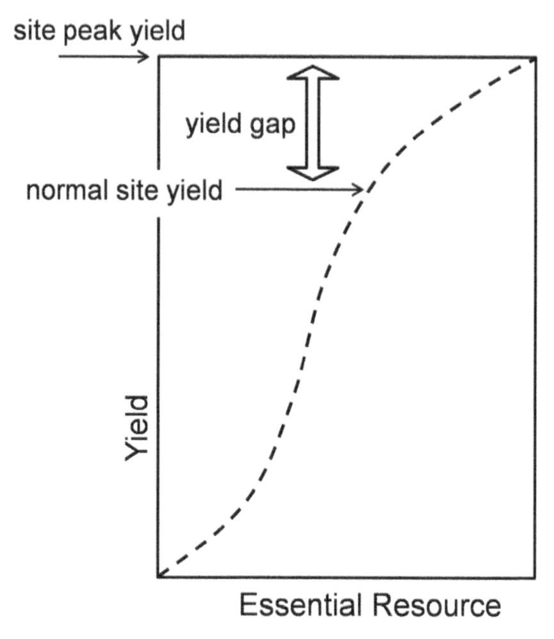

Figure 5.1. A production function illustrating peak and site yields, the yield gap, and their relationship to actual yields.

Resource-Response Functions

A single LER value for a two species intercrop represents four data points. Two of these are the monocultural site yields, two are intercropping data points. With perfect growing conditions, monocultural crop yields would be at a peak or pinnacle. This adds two more points.

These can be the basis for parametrizing general and non-resource-specific yield functions. Illustrating with one-half of the LER equation, Figure 5.1 shows these points functionally presented.

With the standard LER formulation, a value of one represents the site-constrained yields for a single species (1_{site}). Likewise, a value of one can be the ideal or optimal yield (1_{peak}). These points on an input/yield function are show on Figure 5.1. If the curve is labeled such that peak is one, then $1_{site} = Y_a$. It is worth noting that the peak minus the site ($1_{peak} - 1_{site}$) is called the yield gap.

The standard rendering of the LER can be referred to as the site-based LER. For functional analysis, it is more convenient to use the peak as the basis, employ a zero to one scale, and computing the site-based LER accordingly.

As a management tool, both the 2-D and 3-D version (Figures 3.1 and 3.2) of the PPC have planting densities as the x- and y-axes (as in Figure 3.1). There is also a version that has the amount of essential resources as the x- and y-axes.

For this, 1_{site} is changed to a resource-generalized R. With the assumption that both species share the same site and generalized site resources (R), the functional, peak-based LER equation is

$$Y_{LER} = [f(R - P_{ab} + F_{ab})/R] + [f(R - P_{ba} + F_{ba})/R]$$

The purpose is to provide an estimate on what is expected from an intercrop given the amount of site-contained resources. Going forward, there are issues as to how the 2-D, and the 3-D, versions are mathematically expressed.

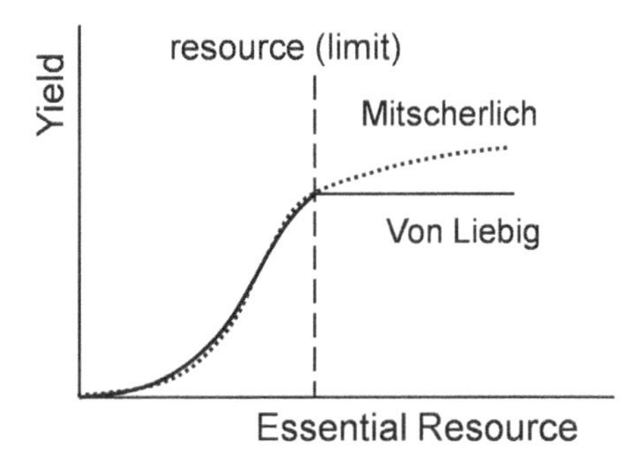

Figure 5.2. The two-dimensional forms for the von Liebig and Mitscherlich, resource-use hypotheses. There can be a number of variations off these hypotheses. Shown is a less traditional sigmoidal form. Their similarity inhibits a statistical determination as to which is best.

Resource-Use Hypotheses

There are differing views on how the essential resources interact to determine plant growth and, ultimately, the yields obtained. This relationship has been debated and the primary governing hypothesis, either the von Liebig, Liebscher, or Mitscherlich, yet to be determined (Paris, 1992). Adding complexity, there are variations on these base hypotheses, e.g., Harmsen (2000), Berek and Helfand (1990).

This may be less an issue than supposed. Despite their conceptional differences, the composite functions are similar in form and are very close in their predictive outcome (Paris, 1992; Llewelyn and Featherstone, 1997). Their similarity is illustrated in Figure 5.2 with variations of the von Liebig and Mitscherlich functions.

Although the von Liebig Hypothesis has found some favor (Grimm et al., 1987), it is more out of convenience, and desire not be become enmeshed in the complexity, that a generalized von Liebig form is utilized in this text. This hypothesis states that the lowest or most limiting essential resources sets an upper limit on yields.

From the previous chapter, single species yields, using a minimum, i.e., von Liebig form, as the constraint, the yield equation is

$$Y = \min [f(L), f(W), f(N)]$$

Broken down, these are, for species a,

$$f(L) = f((R_L - P_{ab} + F_{ab})/R_L)$$
$$f(W) = f((R_W - P_{ab} + F_{ab})/R_W)$$
$$f(N) = f((R_N - P_{ab} + F_{ab})/R_N)$$

Re-expressed, again species a, this is

$$(Y_{ab}/Y_a) = \min [f((R_L - P_{ab} + F_{ab})/R_L), f((R_W - P_{ab} + F_{ab})/R_W), f((R_N - P_{ab} + F_{ab})/R_N)]$$

and, for species b,

$$(Y_{ba}/Y_b) = \min [f((R_L - P_{ba} + F_{ba})/R_L), f((R_W - P_{ba} + F_{ba})/R_W), f((R_N - P_{ba} + F_{ba})/R_N)]$$

where, going back to the base LER equation,

$$Y_{LER} = (Y_{ab}/Y_a) + (Y_{ba}/Y_b).$$

Functional Forms

A mainstay of a monocultural approach involves finding the relationship between essential resources and yields. The norm is to conduct field trials and to select equations as to which offer the best statistical fit to what should be ample yield data. There are a lot of equation forms from which to choose and various criteria from which to narrow the selection (Overman and Scholtz, 2002; Griffin et al., 1987). First is to decide on the correct or a close fit. In general, the greater the number of curve- or surface-adjusting parameters, the closer an equation can be statistically fitted to the data, i.e., more data equals greater functional accuracy.

The problem encountered in agroecology is not at odds with finding and fitting equations to field data, just that the circumstances differ in significant ways. As mentioned, agroecological data is scarce. Hence, it is easier to work with simpler equation forms than the complex versions. The difference lies in the number of parameters. The simplest is the power function. This has one curve-shifting parameter, i.e., $(R)^b$, where b is the sole fitting parameter.

Table 5.1. The four equation forms utilized in this text.

Allometric
$y = ax^b$ = a(x^b) where $0 < x < 1$
Gompertz
$Y = ae^{-be^{-cx}}$ = a exp(–b exp(–cx))
Hossfeld
$Y = \frac{x^c}{(b + \frac{x^c}{a})}$ = (x^c)/(b + ((x^c)/a))
Unnamed
$Y = a(x^b)e^{(-cx)}$ = a(x^b)exp(–cx)

Functional Estimation

In agroecology, three types of curves dominate. These are the (1) sigmoid, (2) power, (3) the exponential growth, and (4) sigmoid with diminishing yield.

A number of these equations that will produce the desired shapes. Those with fewer adjustment parameters (a, b, and c) are easier to fit. The tradeoff is that, with large data sets, the statistical fit might not be as exacting.

The base forms employed in this text are listed in Table 5.1. A more extensive list of candidate equations is found in Appendix 1. Those in Table 5.1 have fewer fitting parameters.

The sigmoid form applies in situations where very little of an essential resource or input produces little yield and, as more is available, yields substantially increase. Eventually, a point is reached where adding more of the resource does not have large impact on output. The Gompertz and the Hossfeld embody this three-phase dynamic.

In other situations, an input has an immediate yield lift. At the upper range, diminishing marginal gains occur. This yield curve is typified by the above-mentioned, outward-sloping power function. For this, the value of the sole parameter is less than one, greater than zero. As an added note, some sources refer to the version in Table 5.1 as the Allometric form. To avoid confusion, this latter term is used in this text.

The third type, that of the inward sloping exponential growth curve, is found where it takes a lot of one input or many plants before significant results are attained. This could often be the case with facilitation. The Allometric function with the exponential greater than one embodies this shape.

The fourth form, has initial high growth and yields, but these diminish as the amount of a resources increased. Although there are a number of forms, Table 5.1 lists an unnamed function (found in Glover, 1957).

Yield Functions

On these functional forms, the research is far from definitive. For what may be the most studied function, that of yield and fertilizer, a seeming majority of the research suggests variations of the Allometric/power function. This is far from an unresolved issue.

Some think the primary yield functions, i.e., light, water and nutrients, as power functions, other prefer the sigmoid form, e.g., Mathews and Hopkins (1999), Hillel (2004), Holford et al. (1992). It is entirely possible that the form does not hold steady, instead, it changes with differences in soils and other variables (Petr et al., 1988).

Decision Range

A key aspect of agricultural, and of agroecological economics, is the concept of marginal gains or marginal returns. Explained through a brief example, the use of one input, say fertilizer, is economically maximized with the following equation.

$$P_i Y' = P_y$$

For this P_i is the price of an input, Y' is the derivative of a generic and unspecific production function, input and yield, with regard to that input, and P_y is the per-unit selling price of the crop.

Under marginal gains, an input is applied as long as it returns a value, in yield terms, that is greater than the cost on the additional unit applied.

A point in reached on the resource/yield function where this no longer happens. This is when the additional unit of input costs more than the value of the additional yield obtained. Managers are not interested in applying inputs when the cost of the additional input no longer increases profits.

It is possible to illustrate using the derivative of the power function where, for the function $Y = x^{0.5}$, the derivative is

$$Y' = 0.5x^{-0.5}$$

Using the 0 to 1 scale, one unit of input (x), when added to soils with 75% of an already available resource, would return approximately 58% of the value of the input. Monetarily expressed, if an added unit of input costs $1 and a per unit of output is worth $0.40 [$1(0.58) > $0.40], this is worthwhile expenditure. If the input costs $1 and the output selling price is $0.60 [$1(0.58) < $0.60], this application will loose money.

This is more than a simple decision tool. Each production equation has, as above, those points that are critical for the decision process. The practical range or reach that contains these points is the decision range. The other, lower sections of the curve are less interesting and, for most agroecosystems, can be of little importance in decision process.

Shown in Figure 5.3, three candidate curves are statistical fitted to the decision range. Due to there closeness, many of the forms would statistically qualify.

The other decision point is zero. Most plants cease to yield or cease to exist before the level of any one resource is exhausted. The lower range (bottom left corner in Figure 5.3) contains assumptions regarding a severe shortfall in resources. It is nice when a function crosses the x-axis at the correct point but, since this is not within the decision range, it is generally not an issue.

The decision range applies to most agroecosystems. It is especially pertinent with revenue-oriented designs. The exception to the importance of the decision range lies with species-complex agrosystems (Chapter 11).

Generic Equations

Given that many of the curves (as in Table 5.1) can adjusted to be close or almost identical within their respective decision ranges (as in Figure 5.3), a generic approach may not be that far afield. It is always better to seek highly accurate modeling with statistically-sound equations. Where data is lacking, slightly less precise modeling with simpler equations may be the best, or only, alternative.

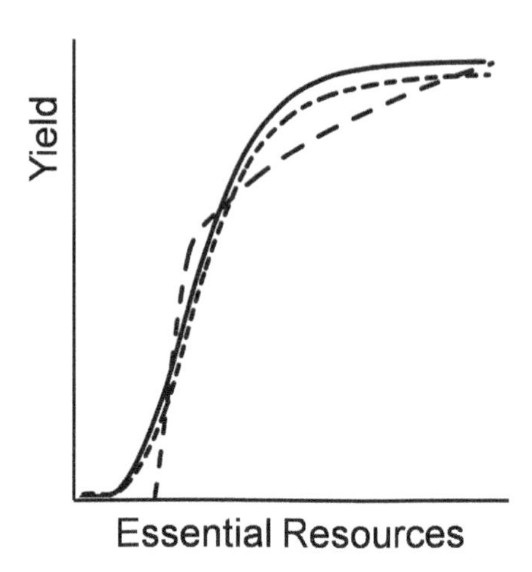

Figure 5.3. Three superimposed production equations. Shown are two sigmoid forms, the Gompertz (solid line) and the Hossfeld (dotted line), and the loser-fitting Allometric function. This illustrates their functional similarity within the decision range.

There are various clues from which to shape these functions. Some of this comes from the relationships presented in the previous chapter. These are roughly summarized in Figure 4.5, p. 66.

In determining peak yields, it should be noted that these values are not a constant, but vary as to the crop variety. For example, many high yielding varieties, including the GM types, surpass local varieties in the obtainable yields on resource-abundant sites. Despite the high peak yields, local varieties may outperform the high yielding types when soils are less than adequate or weather conditions are far from ideal. This implies different resource-response functions.

The site-based LER equation, for the full biculture, six data points. For one species of a biculture, the three points are 1_{peak} along with 1_{site} and (Y_{ab}/Y_a). Irrespective of the limiting resources, the yield gap is often estimated, using actual data, either from experience or past trials. Taken together, these can be the basis for a rough calibration of a resource-response function.

There are a lot of clues, and disagreements, as to the exact form these functions take. As demonstrated in the next chapter, a compilation of the known can go some distance in providing analytical insight.

The Intercrop

Chapter Preview

With the production function, more or less, in place, the functional forms, as they apply to intercropping can be derived. For productive intercrops, this is accomplished by incorporating the components equations, i.e., competitive partitioning and facilitation. The resulting form would encompass a large grouping of agrotechnologies.

Chapter Contents

Introduction

Intercropping is a key manifestation of agroecology with far more species pairings than can be fully studied. The reason for a broad analysis is to deflect the research burden so that economic and ecological understanding is not solely the product of field trials.

The first step is a pre-screening of the commercial possibilities, i.e., which intercrops can be successfully planted and harvested in large or small-scale settings. Beyond that, it would be helpful to have a prediction of the expected yields, best planting densities, and where and under what conditions these might be expected. This is a modeling task where those agrosystems that seem to offer the greatest economic advantage are subject to conformation through limited and specifically-targeted field research.

This chapter looks at the analytical aspects of productive and facilitative intercrops. Out of necessity, not all cropping possibilities (nor agrotechnologies) are presented. For example, species-complex agroecosystems are outside the scope of this chapter.

Productive Intercrops

The basis for productive intercrops is the equatable, or the economically-optimal, distribution of essential resources between two or more closely associated and interacting plant species. This implies that resources are in relative abundance and that competitive partitioning, in all its mechanisms, is the key driver.

The groups of agrotechnologies that fall in this category include

multi-varietal monocultures
simple mixes
strip cropping (seasonal)
barrier or boundary (productive)
treerow alley cropping
strip cropping (mixed tree)
agroforestry intercropping
shade systems (with productive overstory)
multi-species treecrop or forest tree plantations

From an applied prospective, there are, for each, many candidate species and therefore many possible variations. The magnitude of the problem is illustrated by way of the numerous species combinations for a simple, seasonal intercrop (Table 6.1). This list, as long as it is, only touches upon the possibilities. Missing are the three and four species polycultures, many tree-crop mixes of agroforestry, and the tree-tree combinations in plantation forestry.

Economic Orientation

Almost all the intercrops listed in Table 6.1 are revenue oriented. There are cost gains. As some of the management inputs, e.g., land preparation and

Table 6.1. A partial listing of mostly seasonal intercrops [compiled from Vandermeer (1989) and other sources].

asparagus and parsley	lettuce and tomato
asparagus and tomato	maize and cassava
banana and coffee	maize and chili
barley and oats	maize and cotton
basil and pepper	maize and cowpea
bean and beet	maize and cucumber
bean and celery	maize and groundnut (peanut)
bean and cucumber	maize and mungbean
bean and eggplant	maize and muskmellon
bean and tomato	maize and pea
beet and kohlrabi	maize and pigeonpea
beet and onion	maize and pumpkin
cabbage and celery	maize and safflower
cabbage and soybean	maize and soybean
cabbage and tomato	maize and squash
carrot and leek	maize and sugarcane
carrot and onion	maize and sunflower
cassava and banana	maize and sweet potato
cassava and cowpea	mint and radish
cassava and groundnut	muskmelon and radish
cassava and mungbean	oats and pea
cassava and pumpkin	oats and rye
cauliflower and rapeseed	onion and pepper
celery and leek	pigeonpea and sweet potato
celery and tomato	radish and sunflower
cotton and cowpea	rice and sugarbeet
cotton and garlic	rice and tobacco
cotton and groundnut	ryegrass and wheat
cotton and sisal	sorghum and alfalfa
cucumber and radish	sorghum and chickpea
cucumber and tomato	sorghum and groundnut
eggplant and radish	sorghum and groundnuts
garlic and onion	sorghum and maize
garlic and potato	sorghum and millet
garlic and tomato	sorghum and oats
horseradish and potato	soybean and tomato
jocote and pitya	sugarcane and soybean
kohlrabi and lettuce	sugarcane and sunflower
leek and onion	sugarcane and sweet potato
lettuce and onion	tomato and onion
lettuce and pea	tomato and watermelon
lettuce and radish	

weeding, are shared by each component crop, there are, per unit of output, savings. There can be overall, per land area, savings if the intercrop excels at deterring weeds and insects.

Despite these cost reductions, most still qualify as being revenue oriented. Some on the list are designed to be cost oriented.

An example is cabbage raised with a grain crop, e.g., wheat or rye. When growing, the cabbage is able to shove aside the stalks of grain. The grain is present more as weed control measure. Harvesting the cabbage tramples the grain. Later, rather than gathering the wheat or rye, this is allowed a period of recovery and then grazed. This is a cost-oriented system because most of the value derives from controlling weeds, not through the wheat as a pasture crop.

Another example is maize with squash (Photo 4.2). This is a case where economic orientation depends more upon purpose. If the understory squash is planted to suppress weeds, this is cost orientation (also cost facilitation). This is the typical case as the squash harvest is generally low, usually about 10% of the comparative monocrop yields. If dual outputs are sought and the system so managed (e.g., wider spacing), this converts to revenue orientation.

A variation is the maize/bean/squash intercrop. This has the added twist of offering both high yields and reduced per area and per unit costs.

Rules of Intercropping

Deciding which plants intercrop well, and which do not, is often the starting point. For most field work, there are rules for productive intercropping. Originally formulated for tree-on-tree plantations, these have been around for some time. As compiled from Schenck (1904) and Schlich (1910), the four rules are

(1) a light-demanding species can only be mixed with one that is shade tolerant if the light demander grows faster;
(2) a slow growing, light demander is only mixed with a faster, shade-tolerant species if
    (a) provided with help, e.g., pruning or thinning, or
    (b) these are raised in groups, rather than an individual spatial pattern;
(3) when interplanting shade resistant species, the growth rates should be equal or the slower one is protected from being dominated; and
(4) two or more light demanding species should not be mixed, except
    (a) on very fertile, well-watered sites, or
    (b) when the faster growing light demanders are in shorter rotation than the others.

There are some other guidelines. Those species that sequence well generally intercrop well. There are some caveats. With exceptions, water use in intercropping may be a little higher than a sole crop (Szumigalski and Van Acker, 2006; Ofori and Stern, 1987). An intercrop may not succeed on drier sites while a monoculture can yield under these conditions. This is shown with the maize-bean intercrop which, below a certain rainfall level, will not successfully co-plant (Rao, 1986).

There are other sequencing considerations. Since competitive partitioning, and richer soils are the basis for non-facilitative, these systems often immediately postdate a fallow period.

Although rules are inherent in any subsequent analysis, they only go so far. For the simple intercrop, there are other design parameters that must be considered. Among these are the spacings and the planting densities.

Photo 6.1. A tree-crop combination that shows oil palm over cassava. This is an uncommon mix that has not been researched. Because this is tree-over-crop, it falls outside the rules of intercropping. The photo was also taken in Liberia.

Function-Based Analysis

As stated, the rules on intercropping go only so far. In addition to being unable to answer questions on spacings and planting densities, the rules also fall short when dealing with facilitative ecosystems or where there are great differences in the heights of the plants, e.g., where trees are interplanted with crops.

To rectify the shortcomings, one must turn to equation-based analysis. Although development lags, progress is still possible.

In the proceeding chapter, the base LER equation was expanded to include competitive partitioning and facilitation. This resulted in the equation

$$Y_{LER} = [f(R - P_{ab} + F_{ab})/R] + [f(R - P_{ba} + F_{ba})/R].$$

As before, this is based on an unspecified essential resource.

This can be simplified. Changing this is from a site-based LER to peak-based LER ($Y_{P\text{-}LER}$) drops the division by R from the two component parts of the equation.

A key variable for an intercrop are the planting densities. These are optimized by way of the various possibilities curves (production, cost, and/or profit). By adding the number of species to the equation, this becomes both a density-based and a resource-based function.

This requires adding the per area planting densities (n_a and n_b) for two co-planted and interacting plant species. The zero to one scale is reached by dividing the populations by the recommended, per area, monocultural planting density. Restated with this change, the above equation becomes

$$Y_{P\text{-}LER} = f[f(n_a) f\{(R - P_{ab} + F_{ab})\}] + f[f(n_b) f\{(R - P_{ba} + F_{ba})\}]$$

Further postulating non-linearity within the component functions, the LER equation, as a three-dimensional production possibilities curve, becomes

$$Y_{P\text{-}LER} = [f(n_a) f\{R - f(P_{ab}) + f(F_{ab})\}] + [f(n_b) f\{R - f(P_{ba}) + f(F_{ba})\}]$$

For this, the full PPC is the production of four, non-linear functions. Two, competitive partitioning and facilitation, are subsets within the essential resource response functions.

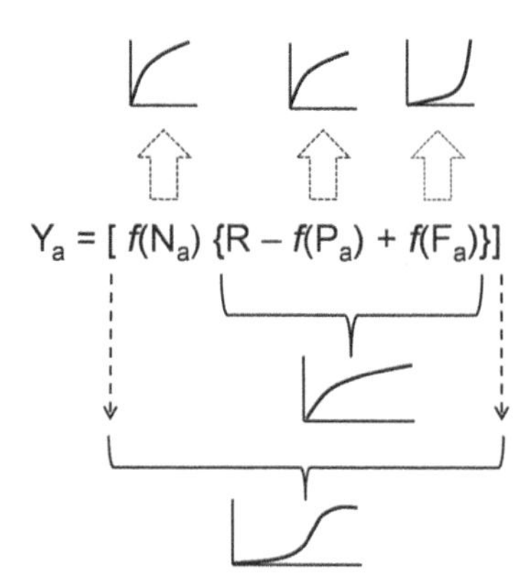

Figure 6.1. The production function broken down into different component curves. This is further explained in the chapter text.

Functional Relationships

In the last chapter, some of the transcendent suppositions are discussed. These being the resource-use hypothesis, e.g., von Liebig, Liebscher, etc., and the form the equations might take, e.g., Gompertz, Hossfeld, etc.

A third uncertainty enters the picture. This involves, not the equation forms, but the mathematical relationships between the variables, e.g., $f(n_a)$ $f(R)$, $f(n_a f(R))$, or some other variation. Although an interpretation is offered here, there can be other relationships especially when going beyond the von Liebig hypothesis.

Component Parts

Figure 6.1 breaks the production equation into component functions, as below. These relationships, as examined in Chapter 4, are the basis for this. The breakdown is, restated, using the LER component for species a,

$$[f(n_a)\{R - f(P_{ab}) + f(F_{ab})\}]_{sigmoid}$$

The sigmoid function is the base. As a monoculture, i.e., where n_a, P_{ab}, and F_{ab} and equal to zero, $Y = f(R)$ is generally sigmoidal. This can be represented by the Gompertz, Hossfeld, or a like function.

Further refinement is possible. Per plant yields, as related to planting density, suggest a power function. In Allometric form, this is

$$Y_a = f(n_a) = (n_a)^{<1}$$

This is a measure on how much of the essential resources are going to each plant based upon that species planting density and the yield that might be expected.

For each species, this is more or less linear until significant plant-on-plant interactions occur. At this point, the function begins to flatten. Representative functions have outward facing slopes. The Allometric function, as above, seems a good approximation.

For competitive partitioning,

$$f(P_{ab}) \simeq (n_b)^{>1}$$

Inter-species competitive partitioning is generally not a factor unless there is sufficient densities of both species. When this occurs, the component species begin to interact. With some, high densities result in severe negative interaction. This is represented with an inward sloping curve and can be approximated by a power function.

Facilitative effects are more complex. For this,

$$f(F_{ab}) = f(R_b)\,f(n_b) = (R_b)^{>1}(n_b)^{<1}$$

Facilitation is greatest when the recipient species (a) is less than adequately supplied with one essential resource. For this to happen, there is a requirement that the facilitating species have a separate source (R_b) for the resource in question and this must be in good supply. There is a need for an adequate population for the facilitating species (b). This varies with the agrotechnology.

For hedgerow alley cropping, a 100% planting of the hedge species would totally exclude the crop (as in Figure 6.3 bottom). The gains occur when about 20% of the area is hedge planted. Above this number (n_b), competitive partitioning kicks in and maize yield diminish.

Other forms of facilitation are optimized when both plant population are near their maximum. This occurs with a nitrogen-fixing or a weed-suppressing ground-cover species. For these systems, competitive partitioning is less, or not, a factor.

There is another possible addition. Since yields are a function of R, this means that $Y_a = f(R)$, and by extension $Y_a = f\{R - f(P_{ab}) + f(F_{ab})\}$, are also be non-linear. The monoculture, in the case, for species a, could provide calibration points. These could be, and have been, e.g., Paris (1992), represented as power functions, e.g., $Y_a = R^{<1}$. The latter is illustrated in Figure 5.2, page 73.

The PPC without Facilitation

For an biculture without facilitation, the PPC is a three-dimensional surface (as in Figure 6.2). With the suggested functions, this becomes

$$Y_{P\text{-}LER} = [(n_a)^{<1}\{R - g(n_b)^{>1}\}^{<1}]_{sigmoid} + [(n_b)^{<1}\{R - h(n_a)^{>1}\}^{<1}]_{sigmoid}$$

The values g and h are the magnitude of the effect. In this equation, they are added only for competitive partitioning. Other such parameters are included as needed.

This equation can produce a multidimensional PPC. Two versions are shown in Figure 6.2. On the bottom, the surface is symmetric in that both species draw equally from the same resource pool. If one species is faster growing and shades the other, the result would be less symmetric form. This is shown on the top.

These surfaces are as one might expect. It should be noted that, for illustrative clarity, the magnitude of each is slightly exaggerated.

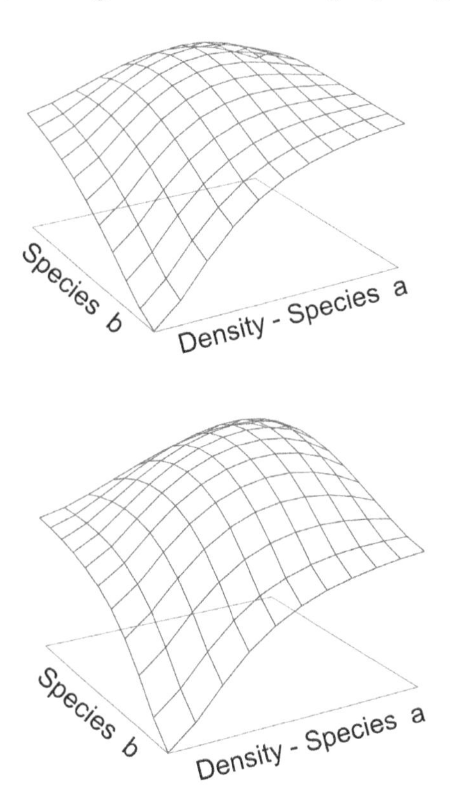

Figure 6.2. A comparison of two intercropping surfaces, one symmetric (lower), the other slightly non-symmetric (upper). Non-symmetry results from the unequal apportionment of one resource, e.g., light as in Photos 6.1 and 6.2. The vertical axis is LER.

Use

This algorithm is tentatively presented with appropriate cautions. For some, this may be a step too far. Still, it is within the stated purpose of advancing a nuanced understanding as to how the parts interact.

Without facilitative effects, an intercrop algorithm would apply to most of the intercrop combinations (as listed in Table 6.1). Once statistically or experimentally solidified, the model, with some variation, could countenance crop pairings other than simple intercrops. This includes the non-facilitative agroforestry combinations, e.g., treerow alley cropping, strip cropping (mixed tree), agroforestry intercropping, and shade systems (with productive overstory).

The uneven division of light, with taller and shorter species, produces an asymmetric PPC. As mentioned, this is within the capacity of the intercrop algorithm.

The same analytical approach can be applied to multi-species forest tree plantations. These are an underutilized forestry agrotechnology with a strong environmental and productive upside.

Photo 6.2. A common intercrop, maize with cabbage, of the type discussed in this chapter. This photo was taken in Liberia.

Facilitative Intercrops

For facilitative systems, the goal is to extract the most from facilitative associations without losses by way of competitive partitioning. The agrotechnologies that incorporate facilitation are

monocultures (with weedy understory)
simple mixes
strip cropping
boundary (non-productive)
isolated tree
parkland
protective barrier
alley cropping (hedgerow)
strip cropping (woody)
crop over tree
physical support systems (vines on or over trees)
shade systems (heavy)

In addition, some of the bio-structures that are in close proximity to a crop, enough so that competitive partitioning occurs would fall under a facilitative algorithm, e.g., with a close-on, anti-insect corridor or with the classic windbreak. The above-mentioned agrotechnologies are described in Appendix 4.

The PPC with Facilitation

Adding facilitation to the previous equation, with some suggested equation forms, this is

$$Y_{P\text{-}LER} = [(n_a)^{<1}\{R - g(n_b)^{>1} + i(R_b)^{>1}(n_b)^{<1}\}]_{sigmoid} + [(n_b)^{<1}\{R - h(n_a)^{>1}]_{sigmoid}$$

The assumption above is for one-way facilitation, i.e., species b aids species a. The parameter i is the magnitude of the facilitation. When the facilitating species does not provide any useful yield, the second species need not be represented separately (extreme the right of the above equation).

The classic example of facilitation is alley cropping where foliage, cut from hedge trees, fertilizes a between-the-hedge crop. Maize being the common crop. Laucaena is a common hedge species. The could result in the a facilitation-based, 3-D LER surface (as exampled in Figure 6.3).

Figure 6.3 shows two theoretical representations that, given the lack of data, are difficult to empirically duplicate. Although not data calculable, these expressions can offer insight.

In the two PPC surfaces, the vertical axes represent the yield of species a, the lower right axes are the amount of a single essential resource available to species a (the per area planting density is held at an optimal), and the lower left axes are the planting density of species b (the facilitating species). The assumption is that density of species b is linearly proportional with amount of a resource added to the system. Finally, the lower corner is the zero origin.

This could example a nitrogen adding system, e.g., laucaena with maize. The upper figure shows the effect of one resource, e.g., nitrogen, without other resources, e.g., light and water. The lower figure is the composite effect. This is the added resource plus plant-on-plant competition for other resources, i.e., $Y_a = \min [f(L), f(W), f(N)]$.

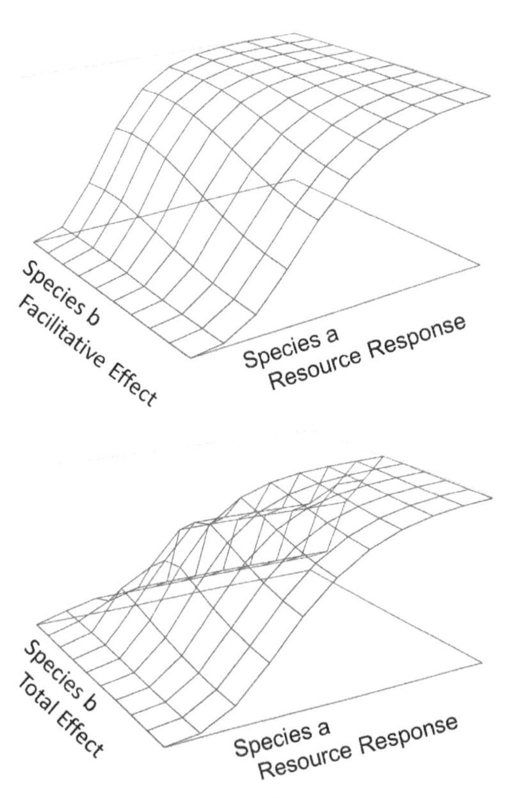

Figure 6.3. In both figures, the lower right axis shows the resources response of the primary species (a) to one resource (the planting density is held constant). In both, the lower left axis is the planting density of and assumed resource input from the facilitating species (b). The vertical axis is the yield of species a. The upper figure looks only at the provided resources sans competition. The lower figure adds competition.

Cost Curves

In cost orientation, the emphasis is on the cost, rather than the revenue side. For a complete picture and where needed, there can be cost analysis that equates with revenue-oriented LER analysis.

From Chapter 2, the cost equivalent ratio is

$$CER = c_a / c_{ab}$$

This can be expanded as

$$CER = c_a / (c_a + c_{ba} - f_{ba})$$

where c_{ba} is the cost of adding the second species, f_{ba} is economic facilitation where, instead of yield gains, costs are reduced. In the crop-with-covercrop example, f_{ba} represents a reduction in weeding costs brought about by a second, purely facilitative species.

This type of analysis also finds use with support systems, e.g., vines on living trellises. Out of necessity, monetary units are employed. Although data can be easier to collection, there is little basis, beyond pure theory, to construct representative cost-based surfaces.

Validation

As stated, a statistical validation of even a small segment of known intercrop combinations will remain an impossibility into the foreseeable future. Those few trials that have looked at planting densities have focused on the two-dimensional PPC, e.g., Ranganathan et al. (1991).

There is another option, one that requires less data and produces a strong semi-validation. This is a statistical fitting data to points along the 50%-50% ratio line. From this, an intercroping competitive profile is obtained.

Figure 6.4 contains four theoretical profiles. These might well represent a hedgerow alley cropping agrotechnology. The assumptions are in line what can be expected in this type of agrosystem. Normally, this is a site severely lacking in one key nutrient, a shortfall that can be supplied by the facilitating hedge.

Based on planting densities, the lower curve (1) is the base function (along the 100%-0 ratio line). The others lie on the 50%-50% line. Functions 1 and 2 show different magnitudes of competitive partitioning. Function 4 has stronger facilitation sans competitive partitioning.

The suggested equations are only starting points. More field-derived insight is needed to insure that the many functional assumptions produced, as in Figure 6.1, are realistic.

Summary

As initially stated, the end goal is to make headway in finding, amongst the large list of possible intercrops, those that have the greatest potential. Carrying this further requires identifying and relating the key essential resources for the different plant species, determining how these interrelate, and deriving and parametrizing functions.

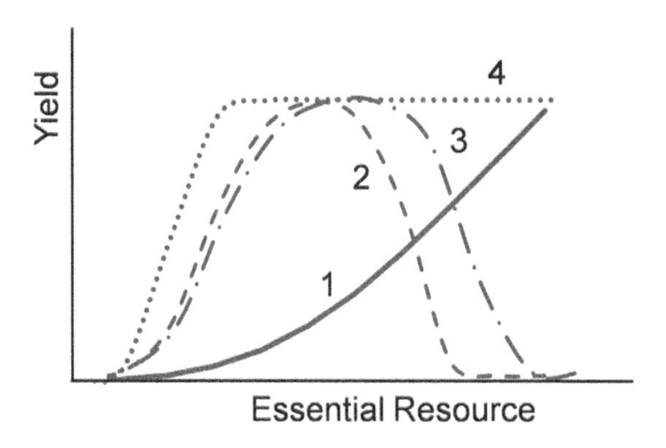

Figure 6.4. The base function (1) along the 100%-0% ratio line and others (2, 3, and 4) that show 50%-50% ratio profiles. The latter three contain differing assumptions regarding the relationship between competitive partitioning and facilitation. This, and the other figures in this chapter, could not be derived without the use of a theoretical model.

This chapter tentatively suggests those functional forms that underlie intercropping. Without more information, the best that can be done are mathematical and visual approximations (as in Figures 6.3 and 6.4) of what is expected.

Given the many uncertainties, the equations presented remains provisional. The theoretical base of agroecology looks at the component ecology. This chapter does this at a numerical, rather than a conceptional level. These may be more a vehicle for understanding than having any immediate in-field application. Questions remain on equation types and their parameters. The idea is to look beyond the numeric barriers and sharpen the understanding on the ecological components interact (as in Figures 6.2 and 6.3).

As an operative corollary with immediate application, it would be advantageous to expand the rules of intercropping. These should include more crop combinations, e.g., trees with crops. Better yet, it would be helpful to have user-friendly, general rules that are able to classify plants as to their innate properties (e.g., essential resource needs) and to couple species accordingly.

Taking a step back, there are other considerations. For revenue-oriented intercrops planted on small farms, there is more latitude and more opportunity for informal trials. For large farms, well-tested combinations are expected. The obstacles lie more in the practical realm, i.e., knowing how to best plant and harvest close-spaced, dissimilar crops.

The situation is different with cost-oriented intercrops. These should be equally suitable for a large or small-farm settings. The problem is less analytical, more in identifying those combinations that offer the sought-after results.

For facilitative systems, the general outlook may be brighter. This is shown with the development of alley cropping for poor farmers. The gains come in not having to resort to costly inputs. For larger farms, the facilitative effects of cover crops may provide a similar answer. It is also a means to address many ancillary issues, e.g., insects, erosion, etc.

As part of the broader picture, the equation form should be, with a bit more work, the basis for optimizing species, spacing, and spatial pattern. Progress requires methodologies that can quantitatively resolve other agronomic questions. The next step forward is a study of the management options and decision vectors.

7

The Agroecological Matrix

Chapter Preview

Agroecology is far more than the intercrop and its variations. For every threat, nature has one or more counters. Full agroecology comes about by exploiting all that nature offers. There are also those threat counters developed outside nature. These are expressed as management options or decision vectors.

All these can be put in matrix form where the cropping threats are rows, the counters are the columns. An important aspect are multipurpose counters, those that address more than one threat. These go a long way in giving agroecology its unique standing.

Chapter Contents

Location
Land Modifications/Bio-Structures
Ex-Farm Inputs
Farm Practice/Environmental Setting
The Functional Matrix

Introduction

In a perfect world crops, crops are grown without imposed limits or threats. Absent an idyllic setting, yields are constrained or even lost. Through the ages, these yield setbacks did not doom agriculture. Now, as in the past, people have found ways to cope with and/or thwart those forces that reduce or negate crops yields.

The threats and their counters are many and varied. Previous chapters have looked at yields, essential resources, intercropping, and a few of the facilitative means to overcome resource-related crop losses. Essential resources are not the only yield-reducing factor. Other threats can change yields for the worst. These too have facilitative counters.

The Cropping Threats

Without productive threats and limitations on the growing of crops, yield gaps disappear. The previous chapters have looked at soil fertility and water. There are always harmful insects and diseases. This do not complete the list. There are others. Roughly categorized, the threats are:

soil fertility (reduced)
rainfall (the extremes; too high or too low)
insects
weeds
pathogens
temperature (again, the extremes)
wind
small animals (birds, mice, etc.)
large animals (deer, elephants, etc.)

Yield Limiting Factors

Economic development (Chapter 3) is based largely on number, as well as the planting densities, of plant species. Held constant, this chapter focuses on yields as limited by a range of threats. Again, starting with the equation

$$Y = \min [f(L), f(W), f(N)]$$

where per area crop yields (Y) are functions of light (L), water (W), and nutrients (N). Broken down and adding internal and two external facilitative effects (F_{x1} and F_{x2}), one component (species a) of a biculture is expressed as

$$Y_{water} = [R_W - f(P_{ab}) + f(F_{ab}) + f(F_{x1}) + f(F_{x2})]$$
$$Y_{nutrients} = [R_N - f(P_{ab}) + f(F_{ab}) + f(F_{x1}) + f(F_{x2})]$$

For practical purpose, there are no facilitative effects with light. A possible exception is the use of microbes to add in-plant shade resistance. Light is included only when in-field biodiversity shades the one or more primary species (and expressed subtractively through the values $-P_{ab}$ and $-P_{ab}$).

The facilitative effects alone produce, in this case, a 2 x 3, mostly additive, facilitation matrix. This is

$$[f(F_{ab}) + f(F_{x1}) + f(F_{x2})]_{water}$$
$$[f(F_{ab}) + f(F_{x1}) + f(F_{x2})]_{nutrients}$$

There are other types of constraints. Some involve an unspecified essential resource. Weeds compete with crops for light, water and nutrients. In these indirect situations, it is not important which resource is limiting as long as the supporting data is valid, the function realistic, and the constraint accurate.

Other cropping threats and limits on yields do not relate to an essential resource. Insects eat and destroy crops without fundamentally altering the amount of resources available to plants. The counters, the F-values, remain the same and/or can be expanded.

It is possible to restate the base equation adding weeds and insects along with other yield limiting factors. This would be

$$Y = \min [f(L), f(W), f(N), f(\text{weeds}), \ldots, f(\text{insects})]$$

In a further modification, this becomes

$$Y = \min [f(L), f(W), f(\text{soil}), f(\text{weeds}), \ldots, f(\text{insects})]$$

These threats, e.g., insects (Y_{insect}), soil nutrients (Y_{soil}), etc., join with the essential resources where one of these functions as the yield-limiting constraint. These can also be expressed as lines in an agroecological matrix.

There are no competitive partitioning losses with insects, weeds, and other non-planned intrusions. When only unspecified mechanisms of facilitation are included, these become

$$Y_{insect} = f(\text{insects}) = [f(F_{ab}) + f(F_{x1}) + f(F_{x2} + \ldots + f(F_{xn}))]_{insect}$$

and

$$Y_{soil} = f(\text{soil}) = [f(F_{ab}) + f(F_{x1}) + f(F_{x2} + \ldots + f(F_{xn}))]_{soil}$$

Adding other threat categories to a yield equation produces

$$Y = \min\,[f(L), f(W), Y_{insect}, Y_{soil}, \ldots, Y_{animals}]$$

Merged with the facilitative effects of water, this can be put into a matrix form as

$$[f(F_{ab}) + f(F_{x1}) + \ldots + f(F_{xn})]_{water}$$
$$[f(F_{ab}) + f(F_{x1}) + \ldots + f(F_{xn})]_{soil}$$
$$[f(F_{ab}) + f(F_{x1}) + \ldots + f(F_{xn})]_{insects}$$

....

....

$$[f(F_{ab}) + f(F_{x1}) + \ldots + f(F_{xn})]_{animals}$$

Threat Counters

Threat counters, i.e., the F-values in the above equation, can be subdivided into single and multipurpose counters. The single purpose counters address one yield threat, e.g., ammonium nitrate mitigates a shortfall of in-soil nitrogen. Multipurpose counters address more that one threat. For example, fallows boost in-soil nutrients while helping to break the cycle of damaging weeds and crop-eating insects.

Both have a role although the starting point for good agroecology should be the multipurpose counters. Only when the multipurpose counters cannot insure sufficient yields, or implementation is the more costly, single-purpose counters find use.

One-On-One Counters

In mono-agriculture, it is common to handle threats one-on-one. Fertilizer, herbicides, insecticides, fungicides, and irrigation each address a single threat. Other single-purpose counters include, against insects, microbes, traps, domestic fowl (e.g., chickens), and introduced predator insects.

Similarly, traps, poisons, hunting, and introduced or encouraged predators eliminate crop-destroying small animals.

Not all threats are totally negative. Weeds can be a positive in regard to insect control. Therefore weed removal, though, for example, hand weeding, would be a negative in the insect line while a positive for soil nutrients.

These one-on-one counters can be included in the agroecological matrix, here as unidentified external inputs G_{x1} to G_{xn}. All are non-linear in that they experience diminishing margin gains as the inputs reach a saturation point.

Multipurpose Counters

A key aspect of agrotechnology is the application of shared ecology. This is where one agrotechnology and/or one decision vector contributes to controlling or defeating more than one threat.

In a comprehensive example, bio-diversity within agrosystems can add to

fertility (nitrogen-fixing species),
experience less soil loss (through erosion control),
conserve rainfall (better capture through increased water filtration),
help control insects (predator-prey dynamics and/or with repellent species),
weeds (suppression through ecological niche crowding), as well as protection from plant pathogens.

As multipurpose counters directly attributable to biodiversity, these are listed in one matrix column (the first column in Table 7.1).

Being able to do more with less translates into cost gains. Whereas the one-on-one counters are inherently costly and inefficient, it is far better to rely upon the multipurpose responses. This is where cost-orientation gains ascendency.

These counters are column-denoted, facilitative values, $F_{ab} + F_{x1} + ... + F_{xn}$. As with one-on-one counters, saturation and diminishing marginal gains indicate non-linearity.

Effectiveness Values

There are two aspects to quantifying the matrix, (1) the functional form each element takes and (2) the effectiveness values. The effectiveness values for one-on-one inputs are, for an unspecified constraint

$$Y = [h_1 f(G_{x1}) + \ldots + h_n f(G_{xn})]$$

where the effectiveness values are, for each counter, h_1 through h_n.

Table 7.1. The Agroecological Matrix. The rankings, good to non-existent, are discussed in Chapter 8.

	Agro-biodiversity	**Bio-diversity**	**Rotations/ Fallows**	**Farm practice**	**Fire**	**Landscape**	**Fertilizer**	**Insecticide**
Insects	good	good	good	fair	good	good	--	good
Soil fertility	fair	fair	good	good	good	--	good	--
Weeds	good	good	fair	good	good	fair	--	--
Wind	good	good	--	good	--	good	--	--
Diseases	good	good	good	good	good	good	--	--
Temperature	good	good	--	fair	--	good	--	--
High rainfall	good	good	good	good	--	good	--	--
Low rainfall	negative	good	--	good	--	good	--	--
Small animals	poor	poor	poor	good	good	--	--	--

As an example, a commercial fertilizer can be 100% effective in providing for the nutrient needs of a crop. For one-on-one, organic sources, e.g., hay mulch, the effectiveness may not be total. Even in large quantity, these may provide only about 50% to 75% of the nutrients that are required.

Adding the multipurpose counters to an unspecified line, with effectiveness values, this becomes

$$Y = [i_1 f(F_{ab}) + i_2 f(F_{x1}) + \ldots + i_n f(F_{xn}) + h_1 f(G_{x1}) + \ldots + h_n f(G_{xn})]$$

where i_1 through i_n are the effective values for the multipurpose counters.

In an undefined line (as expressed above), the different counters are individual production functions. As such, these are mostly non-linear.

Where the entire line is non-linearly defined, it is quite possible that the functions are in some agreement and most of the G-values become linear within the line.

Fallows are examples of non-linearity. These accomplish most of their tasks, e.g., nutrient replenishment, insect and disease control, within the first few seasons or years. There are diminishing gains as the time, and years, extend. For each of these tasks, the fallow is a yield or outward-sloping function. In Allometric form, this would be $(F_x)^u$ where F is the number of years or seasons, u is a number less than one, greater than zero.

For fallows, the effectiveness values relate to time. For the first post-fallow growing season, the fallow is 100% effective in the various tasks. As the years extend, the fallow becomes less effective. By the fourth, post-fallow season, the fallow may be 0% effective. These are the i-values. They could be non-linear although, with so few data points, a linear approximation might well suffice.

Management Options/Decision Vectors

The columns in the matrix correspond to those agrotechnologies and decision vectors being selected. With an eye toward their roles as elements in a decision matrix, the decision vectors are reiterated below.

agrobiodiversity/facilitative biodiversity
genetic/varietal
rotations
fallows
fire
landscape
location

land modifications/bio-structures
ex-plot inputs
environmental setting

Agrobiodiversity/Facilitative Biodiversity

The mixing productive plant species (e.g., an intercrop) to confer a wide range of protections. As previously stated in this and earlier chapters, these can come from productive intercrops or result from plant-on-plant facilitative effects.

In the simple form, the agrotechnologies modeled can be of a preset internal design (with predetermined species and spacings). The design, as pre-formulated, would contribute in hopefully known ways. The advantage is a better handle on yields, revenue, and costs.

Spacings can be influenced by revenue and costs (through the PPC and CPC) and by the site available essential resources. Therefore, the internal design of any agrotechnology may be dependent, more or less, on outside factors.

A more encompassing model combines agroecosystem optimization with cross plot and other external ecological influences. In this version, ecological contributions and the economics of an undetermined agrotechnology design would be less understood and harder to quantify.

Genetic/Varietal Matching

Plants are bred or existing varieties are selected for a number of reasons. A well chosen plant can better resist some threats. These include diseases, herbivore insects, wind, and/or drought. In model/equation form, varieties are mostly a yes or no question. There is some latitude, and some risk-countering gains, for multi-varietal plantings. This would be mathematically expressed by way of the highest expected yield, a slight difference in productive function, and the associated agro-ecology (where applicable).

Rotations

The sequencing of crop species over time to maximize in-soil nutrients and to interrupt the life cycles of detrimental insects, plant pathogens, and weeds.

There are many variations on the fallow theme. As presented in Appendix 4, rotations are agrotechnologies with variations. Because of their overall ecological usefulness, the many contributions are graded as a separate, threat-countering vectors.

Fallows

Allowing the land to idle increases soil fertility while helping to break the life cycle of herbivore insects, some weeds, and most plant diseases. This is temporal facilitation and a column addition. Clearly, time and intensity (number of species and their densities) are variables that, if in abundance, produce diminishing marginal gains, i.e., a non-linear function, possibly in Allometric form.

The main variable is the fallow length. As a form of biodiversity, fallows come in varying forms, e.g., natural, planted, or productive (again, described in Appendix 4). As agrotechnologies, the fallow types can be individual matrix columns that vary on cost, ecology, and yield outcome.

As mentioned, the post-fallow benefits diminish over time. By the fourth growing season, many are essentially naught.

Fire

The burning of crop residues and/or fallow vegetation is a counter to future insect and disease problems and releases plant-contained nutrients. In an agricultural setting, fire is post-season event or the possible end to a fallow. The questions are to burn or not and whether the gains exceed the costs and any associated risk.

As multipurpose addition, intensity may be the main factors in accomplishing the ecological tasks, i.e., the more intense (hotter) the fire, the stronger and more immediate are the benefits. The intensity come through amount of fuel and dryness. The first of these is a function of preceding ecosystem, i.e., generally, the longer the fallow, the more intense the burn.

Landscape

Individual plots are shaped, sized, and positioned to maximize beneficial spillover effects from neighboring plots. The neighboring plots and other, more distant plots can provide protection against wind, high rainfall, temperature extremes along with advancing the predator-prey dynamics that control insects and small animals.

Landscape associations are a major topic with many influences, effects, and practices. Where estimations can be made, each neighboring plot can be functionally expressed with regard to its ecological contributions. More likely is a general landscape ranking, good to bad, for the contribution to each threat counter.

Location

It is better to raise crops on sites there they grow best. Location seeks to do this with minimal site modification and minimal inputs. For plot-level applications, this is considered a given and an analytical constant.

Land Modifications/Bio-Structures

Modifications are physical changes in the land, many of which are directed to better capture available water. Given the wide array of land-modifications, the data situation is uneven and, as a result, many of the structures are difficult to quantify and value.

For many landscape bio-structures, the strength is more a function of distance from the target plot. In the past, there was considerable interest in windbreaks. As a result, there is large body of information on these and like structures, e.g., Caborn (1965) and Prinsley (1992). Their role as an animal barrier is known and 100% effectiveness values are possible. There is also amassed data and determinable effectiveness values for landscape features and insect dynamics, e.g., Altieri and Nicholls (2004).

Terraces (Photo 7.1), as well as contour berms, bunds, and ditches (Photo 12.1), might well merit separate columns in the matrix. The chief

Photo 7.1. For their erosion control and increased infiltration, terraces can be added as a row on the agroecological matrix (Table 7.1). As these are a long-term investment, they would be severely undervalued in seasonal-based matrix analysis. Those in the photo could be hundreds, conceivable thousands, of years old. These are located in Lebanon.

ecological gain in through increased water infiltration and this impacts the rainfall (i.e., retention) row.

Normally, terraces and the like would not be a yea or no decision, whether or not to install. These are long-term investments where analysis decides on the worth of the investment. If already established, the column or columns are needed for a complete agroecological and economic picture.

Ex-Farm Inputs

A mainstay of the green revolution model, i.e., high-yielding crops supplied with ample outside inputs, is revenue orientation. These forgo the cost savings of multipurpose inputs. Most of these originate outside farms, e.g., fertilizers, insecticides, etc. For the positive, this is the most data-rich area of agronomy with well established effective values.

Not all ex-farm inputs are single purpose. Manure adds to soil fertility and can play a role in controlling aphid populations. Manure also helps hold moisture in the soil. In-soil charcoal is another input with multiple gains. Although their primary purpose is passably qualified, there is almost no data on the subsidiary benefits.

As an ex-farm input, one must not forget labor. This is intrinsic with the other columns. With weeds, it represents a stand alone counter.

Farm Practice/Environmental Setting

There are those management inputs that have wide impact, e.g., a tillage methods that interferes with harmful insect and reduces weeds while promoting greater rainfall absorption. Also under this category is the weeding, mowing, use of insect-eating birds, and planting methods. The latter might include, for annual crops, seeds or seedlings. Where trees are involved, this category includes a list of planting options (see Appendix 2).

Many of these are single purpose and each has a matrix column. With little to go on, most are difficult to mathematically define.

The Functional Matrix

The agroecological matrix (Figure 7.1) carries forth into other realms. When formulated as series of production functions, it is an analytical tool. As a measure of agroecological strength, it helps define agroecology (Chapter 13). Also, to be effective, actual numeric values must be substantiated. By default, this converts the matrix into a means to categorize or rank existing knowledge. It also points out where the knowledge base is thin or non-existent.

$$[f(F_{ab}) + f(F_{x1}) + \ldots + f(F_{xn}) + f(G_{x1}) + \cdots + f(G_{xn})]_{\text{water}}$$

$$[f(F_{ab}) + f(F_{x1}) + \ldots + f(F_{xn}) + f(G_{x1}) + \cdots + f(G_{xn})]_{\text{soil}}$$

$$[f(F_{ab}) + f(F_{x1}) + \ldots + f(F_{xn}) + f(G_{x1}) + \cdots + f(G_{xn})]_{\text{insects}}$$

$$\cdots\ldots\ldots\ldots\ldots\ldots\ldots\ldots\ldots\ldots\ldots\ldots$$

$$\cdots\ldots\ldots\ldots\ldots\ldots\ldots\ldots\ldots\ldots\ldots\ldots$$

$$[f(F_{ab}) + f(F_{x1}) + \ldots + f(F_{xn}) + f(G_{x1}) + \cdots + f(G_{xn})]_{\text{animals}}$$

Figure 7.1. The Agroecological Matrix.

There is no one version of the matrix. Each is subject to modification. Where needed, lines are added, others removed. For example, damaging winds come in only a few forms, but insect threats can be subdivided by a broad category and may require consideration down to the level of an individual insect species. In either case, the control of all damaging insects might require many matrix lines.

Columns are also subject to need-based subdivision. Even a single input, such as fertilizer, can be, where needed, broken down into the various elements. In theory, a huge matrix can be formulated that would include all the pertinent cropping information. In practice, matrices would be crop and site specific. They would contain only the information needed for a soil and site, the crops grown, the agroecological options evoked, and the threats anticipated.

Irrespective of size, the matrix is a knowledge framework. This means no missing or weak elements. This leads to suggestions on experimental design where, for each study, there is an active attempt to universally define the results. This means an eye to the broader picture where there is at least one number that can be carried into, and contributes to, a more comprehensive analysis. This should fit within the matrix form.

Published reviews of the literature are, or should be, lines on the matrix. In this role, the better reviews place or suggest effectiveness values for one or more elements on a matrix line.

A Comprehensive Algorithm

Chapter Preview

The agroecological matrix, as presented in the proceeding chapter, can be the basis for non-linear mathematical programming. The objective function being an economic or non-economic expression. With wide application, the resulting objective function and the constraint equations, taken together, constitute a (or the) universal algorithm of agroecology.

Chapter Contents

Introduction

The agroecological matrix is an analytical tool and a means to rank or categorize knowledge. Understanding includes offering a common format for experimental results and, by extension, the information that is sought in field trials. Equally, if not of greater importance, is the role the matrix plays as a (or the) comprehensive analytical algorithm.

In the proceeding chapters, abstract theory has given way to equation-driven, qualified theory. Determining the functional form of each line and each element in the matrix and affixing effectiveness values brings this to the level of more precise quantified theory.

As stated, it is possible to expand the matrix to high degree by adding more lines and more columns. Under insects, these can be lines for individual insect species or categories of insects, e.g., vegetation-consuming beetles, insect-eating flies, etc. It might be better to subdivide soils/nutrition into components, e.g., N, P, K, or by organic content, bulk density, or some other improvement measure.

Threats/Row Constraints

The agronomic threats are many and real. This is a fluid concept as dangers come and go and not all sites face the full range. Restated from the proceeding chapter, the threats, roughly categorized, are

soil fertility
rainfall (the extremes; too high or too low)
weeds
insects
pathogens
temperature extremes
wind
small animals (birds, mice, etc.)
large animals (deer, elephants, etc.)

Soil Fertility

Earlier chapters developed the theory, the equations, and some of the numbers behind on-site soil fertility. This was with both productive and facilitative intercropping. What remains are the ex-plot forms of nutrient enrichment. These include the imported nutrients and those accumulated on-site through time.

The line for soil/nutrients is a mix of one-on-one and multipurpose counters. Soils are thought a non-linear sigmoid function. This is

$$Y_{soil} = f_{(sigmoidal)} [i_1 f(F_{ab}) + i_2 f(F_{x1}) + ... + i_n f(F_{xn}) + h_1 G_{x1} + \ldots + h_n G_{xn}]$$

In Gompertz form, this is

$$Y_{soil} = \exp\{-a_a \exp(-b_a [R_N - i_1(P_{ab}) + i_2 f(F_{ab}) + i_3 f(F_{x1}) + ... + i_n f(F_{xn}) + h_1 G_{x1} + \ldots + h_n G_{xn}])\}$$

The equations parameter a_a and b_a, set the shape of the curve.

It is possible to express soil fertility in one line. It may be better to subdivide fertility into those components, e.g., either N, P, or K, that are thought limiting.

Ex-farm inputs are fairly simple to express. For nitrogen as a separate matrix line, i.e., $Y_{nitrogen}$, $Y_{phosphorus}$, etc., the input of a full amount of ammonium nitrate (a G-value of 100%) would have an effectiveness ranking (*h*-value) of 1.0.

Other inputs many be less effective where the contained nutrient balance falls short in one or more elements. For example, seaweed mulch, with an N, P, K of 1.6; 0.04; 1.7 (Roberts, 1907), might not have the necessary amount of phosphorus. Correspondingly, the effectiveness value for this line element would be low.

There are other column additions. If a plot contains soil-enhancing biodiversity, e.g., nitrogen-fixing plants, an appropriate column would be added, e.g., as $f(F_{ab})$. As previously developed, this would be non-linear.

Rainfall/Water Availability

Crops can be harmed by drought and flooding. Interestingly, these share many of the same counters. All the land modification agrotechnologies have positive effect. These slow and/or capture surface water. The goal is to make water more accessible to crops. This is good during droughts and a way to stop erosion and soil movement when floods happen. In an un-specialized situation, it might be possible to include high and low rainfall on a single line. Some may want to separately consider, and evaluate, drought and flooding.

As with other matrix lines/limits, literature reviews are a starting point for finding, or estimating, effectiveness values, e.g., in regard to water management, Raza et al. (2012). In-plot drought counters include the scattered trees of parkland systems and the water-holding capacity of a high-organic content soil.

As mentioned, terraces and like structures, by way of retention, do add to water availability. Since these are often permanent, they have a set, i.e., fixed, G-value.

Countering the drying effect of wind, often by way of landscape bio-structures, e.g., windbreaks, can be critical. Again, landscape modifications are ex-plot variables, i.e., F_{x1} through F_{xn}.

Weeds

As with other threats, the function relationship, in this case, between weeds and yields is not clear. Some support a power curve, some prefer a sigmoid form (Cousens, 1985).

For the controls and counters, there are some basic options. These start with in-plot biodiversity. This can be intercropping or weed-suppressing covercrop [as $f(F_{ab})$]. The ex-plot options are mostly temporal. Fallows, fire, rotations, and the tillage method, are all viable (expressed as $f(F_{x1})$ through $f(F_{xn})$). There are one-on-one ex-farm inputs. Outside of chemical usage, there is hand weeding and weed-eating ducks.

Biodiversity adds to the options. In many productive and facilitative agrotechnologies, the secondary species (one or more) is tasked with weed control. This may be their primary purpose. This can be through covercrops, above-crop shade, and/or a niche and species diverse intercrop (e.g., Adeyemi et al., 2014). Other control methods rely on tillage, rotations, and/or the post-tillage timing of a planting (Anderson, 2003 and 2005).

The above deals with general weed infestation and responses. This would merit a single line in the matrix. There are weed species that are especially damaging and hard to control. In Africa, speargrass and striga qualify. If special measures are contemplated, then a line for each weed type would be in order.

Insects

Judging from the review material, entomological relationships may be one of the more studied areas of agroecology. An immediate source are comprehensive reviews, e.g., by Altieri and Nicholls (2004).

Many question will remain, especially questions on interaction and coordination of the various controls. Because of this, the overall line is some function of F_{ab} through G_{xn}. F_{ab} through F_{xn} are referred to as Non-Pesticide Management (NPM). The general non-linear matrix line that reads

$$Y_{insect} = f[j_1 f(F_{ab}) + j_2 f(F_{x1}) + ... + j_n f(F_{xn}) + k_1 G_{x1} + \ldots + k_n G_{xn}]$$

where *j* and *k* are insect-related effectiveness values.

Insects are dynamic and populations can explode if not controlled. It is normal to accept crop losses of 5% up to 10% as this keeps a base population of herbivore insects on hand and this insures an adequate diet for residing insect-eating insects. Still, keeping the base populations from expanding puts a higher bar on the effectiveness values associated with any one control.

As with weeds, there are questions on whether to model insect populations, as a whole or to focus on those species that are most damaging. Given the large number of insect types, different lifestyles, and control points, grouping these by the best control would be effective, e.g., those that are controlled by burning crop resides. The idea is to shortcut the analytical process with ready to apply counters and matrix lines that express these. The counters described below are for general insect populations.

Predator/Prey

The battle against herbivorous insects may seem never ending, but it is not without allies. These are the predator insects. To keep the beneficial insects on site, there should be a small, residual population of herbivorous insects. This would entail minor, and acceptable, crop losses (i.e., the above mentioned 5% figure).

The need for a base population places a lower limit on the effectiveness of counters, i.e., all the herbivore types cannot be eliminated. When a predator/prey approach is utilized, the controls are ex-plot, in-plot, and/or close-at-hand landscape bio-structures.

The ex-farm and ex-plot inputs (the G-values) include introduced predator insects, i.e., often those available through commercial sale. Ex-plot inputs are also temporal as with rotations and fallows.

If the continual maintenance of predation types is the main strategy, in-plot and nearby plot habitats are important. These come in different forms and include many types of landscape bio-structures.

All these controls are additive and can be represented, in sum, by an inward-sloping Allometric function, possibility as $(i_1 F_{ab} + i_2 F_{x1} + ... + i_n F_{xn})^u$ where u is a number greater than two.

As mentioned, when predator/prey is utilized, there is always the danger that the population of the herbivore prey can explode beyond what the predator insects can eat. Outbreak protection is part of an insect control strategy. These are generally one-on-one controls. Not all are chemical. These might be the introduction of domestic fowl or a cut-and-carry repellent plant. This could involve adding a second, and a separate, insect outbreak-protection matrix line. This would be if-then type statement that is invoked only when herbivore insect population exceed the recommended base.

Repellent Plant

An alternative to predator/prey is a push with or without a pull. The push is in-plot vegetation that actively repels herbivore insects. The pull is an outside-the-plot plant that attracts the unwanted insects. Usually the pulling plant is located in a landscape bio-structure that is rich in predator insects. Part of the problem is to determine number of insect-facilitative push species that would be needed.

General Expressions

Insects populations and resulting yields are generally modeled in a sigmoid form (Bardner and Fletcher, 1971). This is, with predator/prey dynamics,

$$Y_{insect} = f_{(sigmoidal)}[(j_1 F_{ab} + j_2 F_{x1} + ... + j_n F_{xn})^u + k_1 G_{x1} + \ldots + k_n G_{xn}]$$

As a Gompertz function, this becomes

$$Y_{insect} = \exp\{-a_b \exp(-b_b [(j_1 F_{ab} + j_2 F_{x1} + ... + j_n F_{xn})^u + k_1 G_{x1} + \ldots + k_n G_{xn}])\}$$

Yield-Reducing Pathogens

As with most lines, disease control can be a sigmoidal function. When compared to insects, the base data is less enticing as there are fewer studies from which to estimate effectiveness values.

There are the usual counters, spread blocking bio-structures, in-plot biodiversity (the multi-varietal monoculture or some form of intercropping), in-plot micro-climate, rotations, and the burning of crop residues.

Temperature Extremes

In practice, high temperatures are difficult to deal with as there are few effective counters. Those that do exist involve switching to another productive agrotechnology. These are the shade systems with cooler internal temperatures. As most crops do not thrive in shady environments, this is limited to specifically designed shade systems (as described in Appendix 4).

At the lower end of the scale are frosts. There are counters. Again, shade systems can be effective, but these are not universally suitable. There are landscape designs and ex-plot bio-structures that help. Their effectiveness is a function of location, number, and distance of the control from those plots and plants needing protection.

Wind

Wind can be, and often is, water related, i.e., winds that dry plots and crops. There are also non-transpiration-related effects. Strong winds lodge crops. There is also sandblasting where wind-driven sand particles injure leaves and stems. Although each sand impact is small, the cumulative effect can be significant.

Where winds are light and constant and drying is a chief concern, this line might inclusive with the water constraint. For strong damaging winds, in-plot trees are a facilitative counter, e.g., parkland systems. There are also a series of landscape bio-structures that would be evaluated in this regard.

Small Animals

There are in-plot controls that counter crop-damaging birds, mice, etc. Although overlooked, varieties of grain, by way of awn size, can deter birds (Harrison, 1775). There are facilitative plants that discourage mice, rats, and some other animals (Reilly, 1993). Snakes eat mice and rats and, despite some revulsion, the non-lethal types can be introduced as a one-on-one control.

Habitat can harbor crop-eating birds. However, the same is true of the predator types. Also, there are vegetative counters where select plants repel certain animal pests.

Large Animals

The threats posed by large animals is not usually not a function of a plot or inter-plot design. Hunting and wire fences being the common controls. Where plot design comes into play is where living fences or tree barriers are utilized. These contribute to the overall ecology of plot and would merit a line in a matrix-based model.

Mathematical Programming

Each evaluation is a non-linear mathematical programming problem. Following convention, each contains an objective function and a series of constraints. Table 7.1 is of an applied version of the agroecological matrix with a qualified, subjective, and forward looking ranking of the threat counters.

The Objective Function

Agroecology is guided by economics. By extension, this means an objective function. This can be, and often is, monetarily stated. Other objectives, or combinations thereof, are risk, sustainability, and a series of loosely-positioned environmental goals. None of these are expressed below.

The starting point, the common user goal of profitability, has a linear objective function as

$$\max\ [p\ \{\min\ (Y_{soil}, Y_{water}, Y_{Insect}, \ldots, Y_{animal})\} - \{c_1F_{ab} + c_2F_{x1} + \ldots + c_nG_{xn}\}]$$

where p is selling price(s) of the output(s), the Y values are the individual yields based upon the threats (left column, Table 7.1) and their counters (top row, Table 7.1). The constants, c_1 through c_n, are the costs, implementation and maintenance, associated with each threat counter. These counters can be multipurpose facilitative (as F-values) or one-on-one (as G-values).

Taking the economics out of the objective function, this becomes a hypothesis on how the parts (the existing theories, research, and practices) interrelate. When so modified, the objective reads

$$\min\ (Y_{soil}, Y_{water}, Y_{insect}, \ldots, Y_{animal}).$$

Constraint Equations

The core matrix is as developed. It has, as rows, nutrients, insects, weeds, and water. The multipurpose counters are in-plot biodiversity, fallows, fire, rotations, and the surrounding landscape. These F-values are mostly non-linear. The ex-plot form of biodiversity are the F_x-values.

For the one-on-one counters, there are columns for fertilizers, herbicides, insecticides, and hand weeding. These G-values are mostly linear.

The equations were formulated as suggested in this chapter. The constraint equations for this mathematical programming problem are expresses in sigmoidal form. The matrix elements, alone, are given below.

$$Y_{soil} = f_{(allometric)} [R_s - i_1 f(F_{ab}) + j_1 G_{x1} + \dots + j_{sn} G_{xn}]$$

$$Y_{insect} = f_{(sigmoidal)} [(i_2 F_{ab} + i_3 F_{x1} + \dots + i_n F_{xn})^u + j_1 G_{x1} + \dots + j_n G_{xn}]$$

$$Y_{weeds} = f_{(sigmoidal)} [i_1 f(F_{ab}) + i_4 f(F_{x1}) + \dots + i_n f(F_{xn}) + j_1 G_{x1} + \dots + j_n G_{xn}]$$

$$Y_{water} = f_{(sigmoidal)} [R_w - i_1 f(F_{ab}) + i_4 f(F_{x1}) + \dots + i_n f(F_{xn}) + j_1 G_{x1} + \dots + j_n G_{xn}]$$

The effectiveness values are another matter. Some can be found scattered within the literature. Some are estimated on site. Where numbers are lacking, the qualitative matrix (Table 7.1) can be the basis for a hopefully, close-to-reality reckoning.

Photo 8.1. The sorghum, cassava, yam intercrop analyzed in this chapter.

An Applied Example

In this chapter, a series of interrelated equations are presented. The ultimate purpose is understanding the ecological dimensions of various agronomic expressions. This is demonstrated through a case study.

This example comes from Central Ghana. The photos in this chapter put a practical aspect to the mathematical expressions developed below.

There are a lot of elements with the overall design package. Contained are a fallow, intercropping, rotations, fire, and a favorable landscape where the plots are bordered by uncultivated and/or fallowed grassland.

Agrobiodiversity in the first season (Photo 8.1) comes by way of a yam (*Dioscorea rotunda*), sorghum (*Sorghum bicolor*), cassava (*Manihot esculenta*). There are also a few other minor additions. This mix insures a high initial LER and high total yields.

The gains from fallow and burn extend into second and a third season. Maize (*Zea mays*) is planted in the second year. The final year features low yields of soybean (*Glycine max*) in weed-filled, multi-varietal monoculture (Photo 8.2). No chemical inputs are employed at any time in the cycle.

It is worth noting that there are scattered taller tree within the cultivated plots. These are in the form of a parkland system. Aggregated across the broader landscape, the trees provide ground-level wind protection. As they are widely spaced, they are not a factor in this analysis.

There are a number of mechanisms at play. The pre-planting burn sweeps the area clean of insects, but these restock from the surrounding grassland. The restock would include the predator types. Weeds are naturally eliminated during the long fallow. The post-fallow burn completes the process (Photo 8.3). Most plant diseases are also controlled through these same dynamics. These form the columns or decision variables in this problem-specific agroecological matrix.

Because of the lack of inputs other than labor, this system can be considered cost-oriented. This ecologically-complex, multi-element design is most likely the culmination of long evolution. As such, analysis-based optimization should approximate field results.

This demonstration looks at contributory strength and weakness of the agroecological elements. The design of the internal agroecosystem, i.e., the planting densities, are not in question and not optimized. Only the first year, the highly productive yam, sorghum, cassava intercrop, and the last, the weed-choked soybeans, are modeled. The focus is on the economic interrelationship of the counters.

Results

In theory, multi-year fallows, a clearance fire, and a favorable landscape constitute a potent combination. These should produce an initial cropping period that is weed free, nutrient rich, low in harmful insects, and with few on-site plant pathogens.

Photo 8.2. The weedy soybean monoculture that is the third and last season of the described cropping sequence.

The initial high yields, as an LER value, are estimated at 1.8. From observation and farmer input, this calculation seems accurate. Yields for the final season monocrop, despite the abundant weeds, are about 33% of what could be obtained with a first-season monocrop. This latter number is in line with observations and within a published range (Bolfrey-Arku et al., 2006).

The costs of inputs are critical. The multipurpose counters, being cheaper, are important in arriving at the conclusions. As expected, when these are not employed or there is a rise in cost, the linear program defaults to fertilizers, insecticides, etc.

With this lowest-cost agrosystem, design diagnosis shows the combination of fallows and a clearance fire, when added to the productive potential of the intercrop, are enough to produce the initial high yields. Although the landscape has an ecological role, it is not economically significant when rainfall is more than adequate. Looking deeper, fallows add 0.70 to the LER, the burn adds about 0.10 more. As rainfall declines, the landscape has greater effect on the LER. The fallow remains as the primary yield determinate.

Other insights are useful. For the final season, farmers have choice of increasing yields from 33% to 66% through hand weeding or herbicides. From local observation, many find the trade off financially unrewarding. This may be because they are dealing with the particularly vexing weed speargrass (*Imperata cylindrica*). Analysis shows no economic gain in adding fertilizers in the third and final season. In broad terms, farmers seem to have reached an multi-period optimal solution as confirmed by comparing the observed and this synthesis.

This analysis is not an end all. There are problems on the horizon.

Population pressures are limiting the length of the fallow. The tendency, often encouraged by western experts, is for a solution involving mono-cropping, herbicides, and fertilizers. This solution was looked at and discounted solely on unfavorable economics, i.e., the fallow is more profitable. It might be best to recommend solutions that are less chemical and have the environmental advantages of the current system.

Photo 8.3. The post-fallow/pre-planting burn along with the surrounding landscape.

Suggested Research

In contributing about 70% to the LER and profit, the fallow is the key element. The problem is that the comparatively long fallow (of around seven years) is not tenable with increased land-use pressures. Without the fallow, this design package no longer gives the desired results. This should prompt research interest in the options and alternatives.

As expected, fallows come in various forms. The fallow design utilized here is a burned, woody, natural polyculture. Fire lays the soil bare for planting but, because the tree roots remain in place, this fallow design is mostly limited to hand cultivation. This is less a concern as mounds are constructed for the yams (Photo 8.4).

Fallows control weeds. Certainty in the control of all weeds species, including speargrass, may require a long fallow. Within the fallow ecosystem, the weed-eliminating, broad-leafed, woody plants need time to establish. Once in place, they require additional time to suppress and kill the weed species. These woody plants, cut and dried, provide the fuel for a hot, post-fallow burn. Replicating these conditions is a starting point when designing fallow alternatives.

Halving the fallow period would go a long way to resolving the land-use-pressure problem. Improved fallows with planted woody component are a viable alternative. One option is a planted and burned woody monoculture where success comes in eliminating the grass phase and going directly to the woody component. The trees, usually a fast growing, nitrogen-fixing species, are planted before the cropping period ends. Planting would occur either during the maize or the soybean phase. This, along with the overall effectiveness, are research questions. Ideally, the maize or soybean crop would benefit, either through increased yield or better weed control.

The algorithm suggests that this could happen but, at this stage, it is not known how competitive the trees would be with the latter-year soybean crop. They could help by suppressing some weed species or hinder by magnifying the on-crop competition.

This is also a more costly option than a natural fallow. The tree planting method could be important. A less expensive method, such as branch planting, would make or break this alternative.

There are other fallow designs that are further afield. One is a burned, non-woody monoculture that employs a vine species. This is similar to tree fallow except an aggressive and blanketing vine species is employed. The hope is that this would offer the same degree of the weed control as a woody fallow.

There are other options that alter the current system. These include some undefined mix of alley and intercropping. This can be interrupted with an occasional, brief fallow.

A quick analytical look at this option shows that net income could more than double than is possible with a chemical-supported monoculture. This combination seems to offer a high degree of design flexibility. Still, this not an easy fix as this is a major change from current practice. There agronomic questions, e.g., on intercropping within an alley, that are outside the predictive parameters of the algorithm as demonstrated.

For the other fallow options, there are other issues that must be investigated. These are the social dimensions and the impact on long-term land rights. Even without all the design specifics, as is the case here, the analytical results provide a preview of the projected profitability. They also foretell the critical ecology.

Photo 8.4. The construction of mounds for the yam is part of the pre-planting, post-burn site preparation.

Gains

With the ability to piece together and evaluate complex agrosystems of any form, the analysis must deal with multiple combination and a severe data shortfall. In the example above, the algorithm shortcuts, or a least previews, what would normally be a farmer undertaken, often lengthy, agroecosystem evolution.

At the fringe, the model can estimate the economically optimal amounts for the various agrochemicals. However, more is expected. Lacking ecologically quantified breakdowns of complex agroecosystems, comparisons must be made against existing examples or qualitative descriptions. The assumption is that local farmers have succeeded in arriving at the best of many possible designs. When this happens, the results should agree.

Better ecological and economic interpretation allows for better decision making. Highlighting the most contributory ecological relationships greatly reduces the improvement process. At the same time, it avoids wasting effort on those dynamics that add little to the profit and/or productivity.

Through this model, initial on-site observations and recommendations are confirmed or challenged. Because the model-based second opinion is optimized, this removes much of the speculation as to the end point of an evolutionary convergence. Challenges, rather than being off the mark, may portend the developmental future of the agrosystem in question.

In economics, investments that offer the greatest return are the first considered. This holds true for research. From the analysis, users know which threat counters give the greatest return. The returns from research are determined, first through the increase in the appropriate effectiveness value(s) and then, after adding selling price(s) and costs, through the change in projected profit. Multiplied by the total cropping area and the number of farmers using the agrosystem, one has a good estimate of the broader gains.

Time, experience, and specific examples will be required, both to explore the full range of applications and to introduce users to the notion that sustainable agriculture can be approached by way of full-on agro-complexity, as in the example offered here.

Presenting the matrix expanded to equation form as a (or the) central theorem of agroecology may be an overreach. Still, a core theorem or an underlying set of equations could well exist. The existence of underlying base equations and the form these take should be part of an agroecology give-and-take.

There may be a mathematical climax that represents the common cropping situations. Agroecology does not end here. There are other topics that feed into and provide a more complete picture.

Temporal Agroecosystems

Chapter Preview

The agroecological matrix, as non-linear programming form, is powerful analytical tool. In conjunction with the equations for intercrops and like systems, they cover many of the cropping possibilities. Despite the universality of these algorithms, there are specialized systems that require a second look. Temporally-based agrotechnologies are part of an extended analysis. Some of this involves a change in the objective function of the matrix approach and/or the use of multi-stage non-linear programming. Much of the focus here is on taungya-type agrosystems.

Chapter Contents

Introduction

Within a classification of agrotechnologies, there are those where biodiversity is expressed in both space and time. These are the temporal agrotechnologies. These came in varying forms.

The previous chapter analyzed a Ghanian yam, sorghum, cassava intercrop. The first-season intercrop was followed by second-season maize. In the third and final year, soybean was the sole crop. This sequence is post-fallow dependent as the intercrop is the most productive and most valuable. The subsequent years are less of an economic necessity. Minor yields and profit result by exploiting the nutrients that remain after an intense, post-fallow cultivation.

Although this example shows a number of different temporal strategy, there are others. The overall purpose is to take advantage of the ecological dynamics that accrue over time.

Agrotechnologies

The temporal agrotechnologies, listed, are

- Fallows
    - Natural
    - Planted
    - Productive
- Rotations
    - Single rotations
    - Series rotations
    - Overlapping cycles
- Taungyas
    - Simple
    - Extended
    - Multi-stage
    - End stage
- Continual

Briefly, a fallow can be natural, allowing vegetation to reestablish without interference. Planted fallows are just that, vegetation is planted for specific purpose or to shorten an otherwise long fallow period. Productive fallows are an option. These allow some minor outputs, e.g., grazing, while the land regenerates.

Planting the same crop in each growing period is not considered a rotation. The same is true when the crop species seasonally changes but, lacking inter-seasonal planning, this would not always ecologically qualify as a rotation.

A series rotation is the most common expression. This spans many planting seasons where crops are grown in a specific order. The purpose is to maximize yields from season to season. There are a number of gains from more available nutrients, to better in-soil moisture, better weed control, etc.

Overlapping cycles are rather rare and mostly confined to the tropics. This is where some crops are planted after others have been in the ground for some months. These are found in tropical garden settings where continued year-long harvests are the goal. These are also found with mixed treecrop systems, i.e., multi-species orchards.

Taungyas are temporal sequences associated with forest tree or treecrop plantations. These have wide geographical use and generally operate without mainstream assessment, e.g., King (1968).

Taungyas start when a plantation or orchard is established. For the simple taungya, crops are planted to exploit the space and under-used essential resources available when the trees are small. In addition to another yield, the crop helps suppress weeds, can decrease the erosion threat, and/or can contribute in other ways. Cropping ends when the trees reach a set height.

The extended taungya keeps the crops for most or the entire tree cycle. The crops usually change to match the changing growth conditions. For example, more shade resistant crops are established as the understory light diminishes.

Multi-stage taungyas have, in essence, two primary species. After the first set of trees has been in place for years or decades, another yielding tree species is planted. These coexist from many years, after which, the first planted is the first harvested. After the first harvest, the second planted becomes the sole primary species. A multi-stage taungya may also include crop understory.

The end-stage taunga occurs with tree plantation where a some trees are harvested before the full cycle ends. This allows those remaining to grow more rapidly, to greater volume, and to a greater value. The space and surplus resources can be use for a crop or, more commonly, as pasture.

Continual systems maintain the same agrotechnology in perpetuity. Usually these are some forest or agroforest-type system (Chapter 11). Although the agrosystem in place for a long period, the actual species mix could gradually change over time. This change in the output mix might have an economic motive.

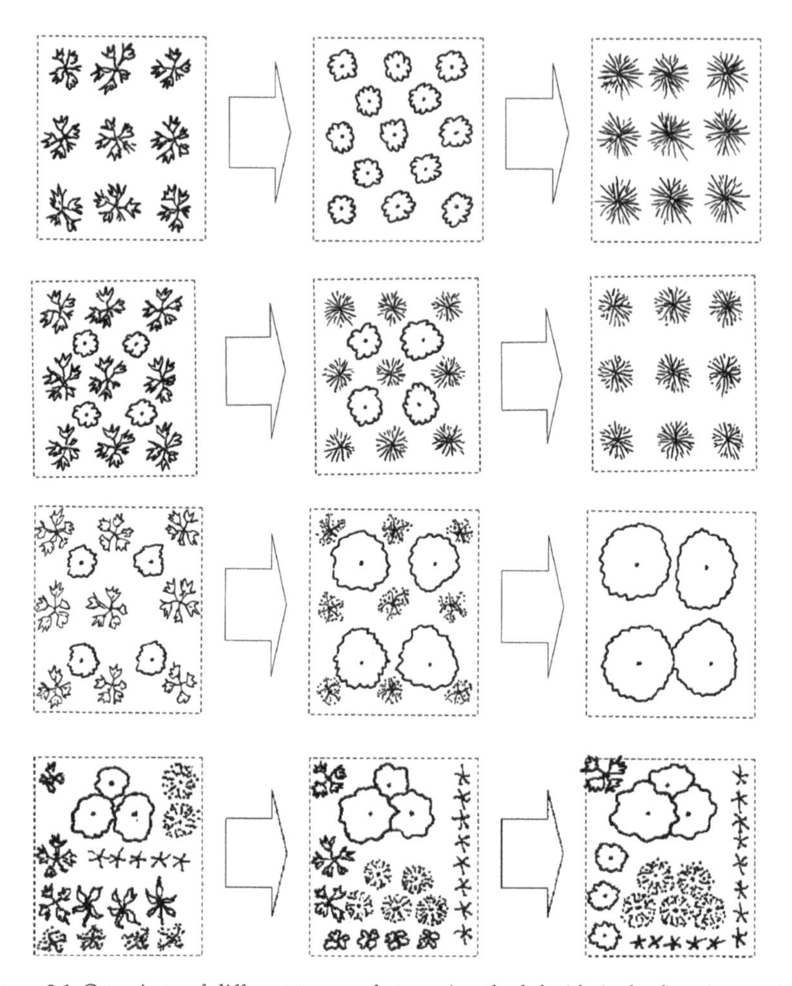

Figure 9.1. Overviews of different temporal strategies, the left side is the first time period, the right is a latter period. From top to bottom, these are (1) series rotations, (2) overlapping cycles, (3) a simple taungya, and (4) a continual agrosystem.

The classifications, as described above, are demonstrated in Figure 9.1. This shows, from top to bottom, series rotations, overlapping cycles, a simple taungya, and a continuous system.

Ecology

Inter-seasonal bio-diversity brings about potentially favorable inter-seasonal ecology. This can confer long-term sustainability allowing for continued cropping. This might not always bring about optimal soil fertility and optimize yields. It could serve to lessen the amount of fertilizers and other applied nutrients.

Nutrients are not the only gain. A long-term strategy applies to damaging insects and to the suppression of plant diseases and weeds. The idea is to break or interrupt the reproductive cycle of the various destructive organisms. This makes these, as future problems, easier to manage.

When applied as a disease control measure, longer seems better. Examples are plentiful. A period of six years eliminates nematodes and rhizoctonia in potato and prevents carrot leaf spot and wheat head blight. Some crops, e.g., carrots, even longer rotations are needed to eliminate some key pests (Reinders, 2007).

Another potential gains is through in-soil moisture. Rotations can be formulated so that organic content of the soil remain high and more water is retained and available.

Examples

In accomplishing the above, there are long-established rotational partners. Wheat should follow beans, vetches, and clover. Barley or oats should come after turnips, carrots, or potatoes (Rham, 1853).

Others are soil specific. For poorer sites, a fallow is followed by wheat, then grasses, oats, peas or beans and, to complete this six-year sequence, wheat. For rich soils, a fallow is followed by wheat, then beans and again wheat. A fallow restarts the sequence (Loudon, 1826). For the above, wheat might be considered the primary crop.

Other examples of series rotations (also Rham, 1853) are

maize --> bean --> potato
tomato --> onion --> squash
oats --> clover --> ryegrass
soybean --> groundpea (peanut) --> sunflower
wheat --> clover --> alfalfa --> ryegrass
clover --> alfalfa --> ryegrass

Economics

The economic potential lies with two temporal agrotechnologies, the series rotation and the extended taungya. The series rotation should be part of all intensive cropping situations. This applies to a large percentage of world agriculture.

The taungya use is more limited. Again, there are commercial applications mostly with treecrop and forest-tree plantations. In some tropical regions, these are many in number and large in area. The extended taungya countenances food production and, if well done, increased profitability from these plantations.

The taungya emphasizes the tree or treecrop, not the understory species. Competitive partitioning is not part of this relationship as the understory should not be a negative against the primary tree species. Species selection, timing and spacing should insures that this does not happen.

There can be complex economic relationship between the trees and the crop. On the surface, there is the added revenue of a second crop. There are also the facilitative gains where the trees benefit from the crop. The latter comes through weed and erosion control. Less observed is when fertilization, applied to the crop, is partially captured by the trees. There is also the increased monitoring of the trees that is a direct corollary of a crop presence.

Photo 9.1. A first year taungya showing popular trees over sugar beets. The tall trees are the result of stem planting (described in Appendix 2).

Multi-Participation

Often associated with temporal agroecology is multi-participation. This is where two or more landusers work and harvest from the same plot of land.

This can be simultaneous or sequential. The work and harvests are usually divided as to the crop species.

With divergent user goals, this has limited application but, where the socio-economics are favorable, this can promote specific agrotechnologies. The historic record has been with the simple taungya.

In the first year, farmers prepare the land and establish a crop. At the same time, trees are planted. The trees coexist with the crop and, if there is enough free space around each tree, growth is not impacted. The gains are mutual, the farmer harvests and sells a crop while those that own and manage the trees do not incur land preparation costs and, in the first and most weed-susceptible year, weeding costs.

Beyond this, there are other simultaneous, multi-participant possibilities. For large-scale producers of crops, trees, and treecrops, the open land between the plants can be similarly sharecropped. This requires nearby land—poor farmers who are seeking an opportunity to farm, even under somewhat restrictive conditions.

Multi-participation can be inter-farm rotational. This is where land-users specialize in one crop and raise this one crop on different farms. This would be part of a planned rotational sequence where each participating landuser has the one crop on at least one farm. The notion is to promote crop rotations.

Series Analysis

In finance, the customary evaluation methods are based around Net Present Value (NPV). This finds application in agroecology and is a key measure in commercial farming when loans or investments are involved. As design tool, NPV has two drawbacks, the first is the currency unit, the second is the need for a discount rate.

It is possible to discount the LER or RVT. Five years of LER values, e.g., 1.12, 1.10, 1.0, 0.97, and 0.94, discounted at 4%, results an NPV of 4.58. Rather than cloud the analysis with a discount rate, it is easier to add the seasonal LER, RVT and/or CER values and compare these against the values for the alternative cropping sequences.

In another hypothetical example, comparing the LER values for two rotational patterns. For species a through e over a five years period, this is

$$(LER)_a + (LER)_b + (LER)_c + (LER)_d + (LER)_e = 5.3$$

$$(LER)_b + (LER)_e + (LER)_c + (LER)_a + (LER)_d = 4.9$$

Clearly, the first rotational sequence is the better. This method more than suffices for series rotations, not so with the taungya. Series rotations can occur with intercrops and, as soil conditions change, so might the subsequent intercrop mixes. With some soil conditions, monocrops might prove the best option.

The optimization of series rotations are more soil related than a pure economic problem. The economic input comes is choosing which of the alternatives is best. In one example (Rham, 1853), i.e., where barley or oats is planted after turnips, carrots, or potatoes, there are 10 combinations from which to choose.

This rotation was suggested in the pre-fertilizer era. Fertilizers reduce the need for fallow-obtained nutrients. When this happens, a reordered sequence still offers advantages. These include pest and weed control.

In theory, costs are better controlled with a recognized sequence as a starting point. The idea being that a less costly fertilizer or nutrient source can be utilized or one application may be sufficient for more than one growing season. This could reduce fertilizer runoff.

There are other options. Fertilizer could shorten the fallow. This depends on the importance of the fallow in controlling pests and weeds. These are open questions.

Time Area Equivalent Ratio

There is a ratio designed to measure the gains from temporally uneven intercrops. From Hiebsch and McCollum (1987), this is the Time Area Equivalent Ratio (TAER). For a two species intercrop, the equation is

$$TAER = \{[(Y_{ab}/Y_a)(t_{ab}) + (Y_{ba}/Y_b)(t_{ba})]/t_t\} + 1$$

where t_{ab} is the time it takes to intercrop species a with species b, t_{ba} is the time is takes to raise species a with species b. The total time period for the intercrop is t_t.

This ratio gains prominence with long-period agrosystems, specifically the taungyas. For a simple taungya, the above equation would have species a as the long-duration tree, species b the single season crop. For this, $t_{ab} = t_t$ expressed in years. The number one above represents the yield of the primary species.

With a 30-year tree rotation, the expected TAER for a simple or for an end-stage taungya would be low, about 1.02. In contrast, the TAER for an extended taungya with a similar rotation, the TAER would be about 1.70.

The TAER is somewhat limited as a shortening of the overall rotation represents an economic gain that is not always reflected in the TAER. Despite this, the TAER has some value in comparing and optimizing crop sequences for extended taungyas. It is also useful looking at the intricacies of the multi-stage taungya or for less biodiverse versions of the continual agrosystems.

Photo 9.2. A second or third year taungya showing popular trees grown over oats. This and the first photo are from Chile.

Taungyas

Having one or two tree species that span the full temporal cycle adds complexity, but does anchor the analysis. There is an optimal sequence that includes the length of taungya cycle and the types and durations of any included seasonal crops.

Rules for Taungyas

For both simple and extended taungyas, there are some obvious rules,

(1) in each year an under-tree crop species is planted that will provide the most profitable yields given the growing conditions as set by the primary (tree) species, and
(2) the crop should be spaced and managed so as to not overly affect overstory tree growth and/or yields.

There are assumptions connected with these rules. Foremost is that the trees provide a far more valuable harvest than the crops. This may be the case in commercial settings where industry relies on a set in-flow of raw materials, e.g., latex from rubber plantations, wood from forest-tree plantations.

This is why even a simple pastoral system, grazing under trees, may not find favor. Although grazing is enticing as a low-cost method to control weeds, pasture animals can eat the foliage and step on young trees. As the trees grow, animals can still gnaw on and damage the bark. A well-controlled, simple taungya may be only option.

This would not be the case for small farms where the most profitable mix may result in more understory crops and a reduction in tree growth and/or yields. The above rules are modified accordingly.

Taungya Analysis

Depending on the user, simple or extended taungyas can be reduced to a series of seasonally discrete decisions. Farmers with small land holdings and some commercial growers may be disposed to an economic tradeoff between the crop and the tree.

The larger plantations, those faced with set quotas, may not be willing to delay seasonal growth or seasonal yields. For tree plantations, any increase in the rotation time in order to get to a set wood volume or tree diameter is a lost opportunity.

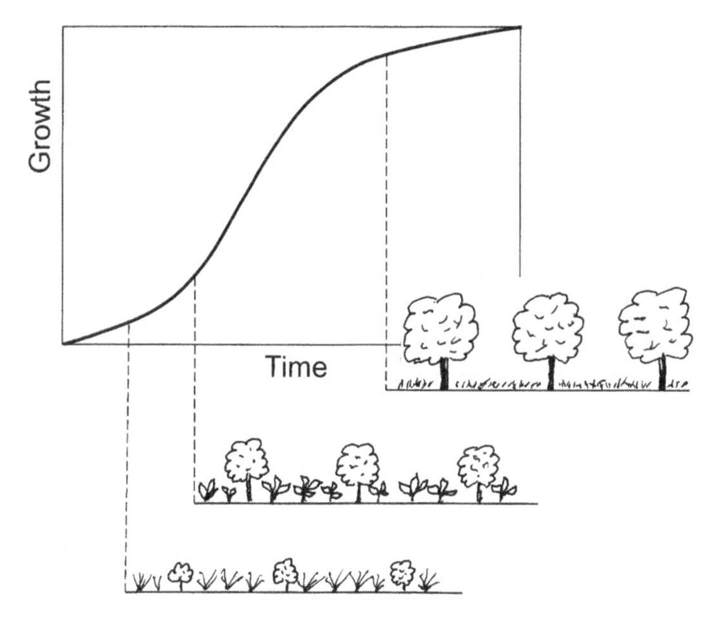

Figure 9.2. An extended taungya showing how the production function for the trees relates to the different temporal stages.

The opportunity costs come, for example, when a delay of five years in 25 year tree rotation (i.e., going from 25 to 30 years) means a delay in the start of the new plantation. Two costs are incurred (1) obtaining the wood five years after expected and (2) a postponement in the first few, and the more productive, taungya crops. These must be weighed against often low productivity of any latter-year, below-the-trees cropping activity.

For high yielding plantations or orchards, those that truly maximize productivity, will make full use of the extended taungya. Rather that being relegated to margins lands, these are instead situated on prime farmland.

Going from a simple plantation without a secondary species to an extended taungyas completely changes the situation. The first impact is with location. Given the higher possible yields and potential increases in profitability, the plantation may now be situated on prime, irrigated farmland rather than on a hillside.

With a better site, other options enter the picture. This included tree-planting where stem planting, rather than seedlings, are employed (see Appendix 2). If the stems are long enough, these can cut 10 years off the rotation. This also contributes to highly improved tree quality. This is because only the best stems are selected for transporting (as in Photos 9.1 and 9.2).

The taungya sequence begins with various first and second years crops. If there is no tree thinning and no end-stage taungya crop, these sequences almost always end as a grazed pasture-beneath-tree system.

Photo 9.3. An early stage rubber plantation showing a lost opportunity. With this, there is plenty of open ground, space for a taungya crop, and the possibility for early income. This could be a multi-participation effort. Older rubber plantations might include shade-resistant, understory crops such as coffee and cocoa.

Summary

Planned sequences should be the norm for seasonal cropping. This is an established agrotechnology where the history of use, and known successions, reaches back to antiquity. There are plenty of starting points and, from an economic perspective, rotations do not require sophisticated analysis, the gains should be apparent.

Taungyas have shorter history. As a studied agrotechnology, these go back to the mid-1800's. Generally, these are underutilized. For forest-tree plantations, increased use would benefit by promoting a shift to those tree species that are intercrop favorable, e.g., changing from pines to poplar. This is not likely to happen solely for intercropping convenience. Consumer preference would be the best force for change.

For treecrop plantations, the path to increased taungya use is less fraught. There are only knowledge and acceptance barriers. It is a more difficult sell with treecrop plantations or large orchards. Most do not embrace this as a first stage or as a life-of-the-plantation alternative. These is no economic methodology that would be helpful in acceptance. Successful demonstration trials with the major treecrops would be the best step forward.

10

Risk Assessment

Chapter Preview

At a time when climate change and associated risk is at the fore, risk considerations has gained in importance. This chapter looks mostly at the more practical, seldom-enumerated, farm-level, anti-risk strategies. Since many of the anti-risk measures are barely mentioned in the literature, e.g., disarray and scattering, related evaluation remains rudimentary.

Chapter Contents

Introduction

Agronomic risk can take many forms. Too many populations experience varying degrees of social or political upheaval, e.g., war or revolution. These disrupt the availability of food. Even faraway strife that interrupt the distribution of land-grown products or of those inputs that local agriculture needs. This type of risk is measured by way of sustainability and the output/input ratio (Chapter 3). Despite their importance, social upheavals and their agro-consequences are not included in this chapter.

Crops are lost in other ways. Plant-eating bugs can destroy yields and plant diseases are always looming. Since agroecology is an economic activity, low crop selling prices can be a threat. Price-caused bankruptcy does not compare with starvation but, for a farmer near the financial edge, this is still serious.

The biggest long-term threat to land-produced fuel, fuel, and fiber may be climate change. This can have a directly impact or aggravate an already worsening situation. This is a big unknown. There are plenty of reports, from leading governmental and non-governmental agencies, predicting a grim future. This may not be a dire as some might say. In previous chapters, risk was mentioned. Countering agro-risk is a strength of agroecology.

Always importance, climate change is bringing risk, as a topic, more to the fore. As a future threat. In a few cases, a change in weather patterns can be for the good, e.g., additional rainfall in a dry climate. This is far from universal. The predictions are that, in most regions, rainfall will lessen or become erratic. This might be accompanied by higher average or higher maximum temperatures, higher transpiration rates, and less in-soil moisture. This makes risk-reduction a greater priority, if not a necessity. For this reason, it is opportune to look at the theory, the counters, and the mathematics of risk.

There are some that see climatic variability on a daily or seasonal basis. This is where fluctuations in temperature and/or rainfall are normal and expected. Without built-in countermeasures, crop yields would be reduced or lost. Those residing in these marginal zones have developed suitable cropping systems. Study of these systems is generally relegated to the fringes of agronomy and, when not studied, the techniques can be lost.

A complete listing of all the risk-countering strategies is not possible in a single chapter. Instead, this chapter offers an overview of some of the broad anti-risk strategies.

Economics

When climate intercedes, the economic priorities often change. Profits can be the number one goal when famine or starvation are not a seasonal or a distant worry. Those operating in a stable climate, those with ample irrigation water, or those with crop insurance and/or governmental protections are less at risk. Millions of farmers do not have these protections and are exposed to climate-related or other forms of adversity. Where this happens, food security can trump profits as the primary farm objective.

Basic Risk Counters

Weather and climate-change related risk may be the most damaging and receive the most attention. Most is water related. Droughts can dry crops and reduce yields. Figure 10.1 diagrammatically shows the dramatic yield plunge that can occur when a resource (e.g., water availability) is on the edge.

Also possible, very high precipitation can be an equal threat where saturated soils can retard growth. These losses do not occur with all crops. Tuber crops can be water-saturation sensitive. This situation is shown in Figure 4.4, p. 64 where soil drying or less moisture can result in yield gains.

With more rain, erosion becomes a concern. Also, the moist, humid conditions can accelerate plant diseases.

It should be noted that many of the same mechanisms that insure yields in good times also insure yields when adversity threatens, i.e., infiltration ditches help mitigate short periods of low rainfall while also protecting against sudden and high precipitation. It is difficult to assign precise matrix element numbers, but the effect is time-tested.

There are many techniques to manage productive uncertainty. Commonly, this is done by defending plots through inputs and placement. The latter comes by way of known threat counters and associated agrotechnologies.

Ex-Farm Inputs

Many view the defense of crops as single mechanism, e.g., spraying against insects, irrigation to counter droughts. These constitute single element, one-on-one defenses and are singular values in a matrix line (Chapter 7).

Irrigation systems with large water sources or reserves can, as expected, help bridge a low rainfall period or even protect against shorter droughts. There are other benefits. The water put onto a field, through flooding or

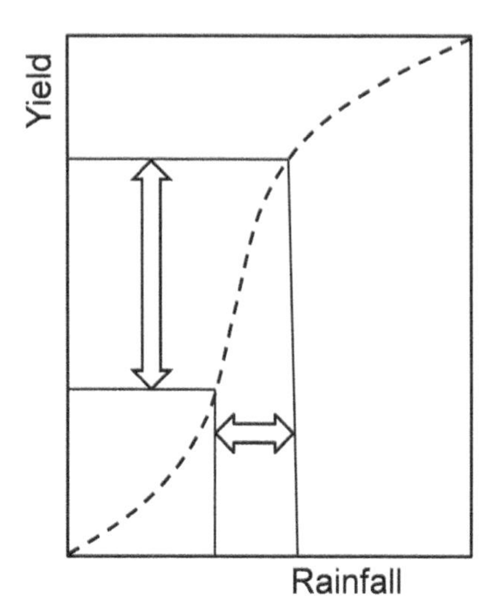

Figure 10.1. The scenario where slight more rainfall, less evapo-transpiration, and/or more in-soil moisture dramatically increases yields.

sprinklers, can counter extremes in temperate, large amounts of water can cool crop when it is too warm and warm crops when they becomes too cold. These irrigation-based counters function best when the extremes are brief and relatively infrequent.

Farmers turn to irrigation to increase yields. Related, the purpose can be to grow risky crops on marginal sites. In doing so, this can magnify the uncertainty. The better option may be to employ multiple defenses. Often these allow yields on risky sites without irrigation.

Land-Modification/Bio-Structures

There are the land-modification agrotechnologies that can boost available water and prevent a yield plunge. Contour ditches function by allowing for more water infiltration and retention for those rains that do come. There are bio-structures that do much the same. In addition, these can block the wind and slow soil drying. The land-modification agrotechnologies are listed in the left column. Those that are vegetative are listed on the right. All are described in Appendix 4.

Absorption zones	
Microcatchments	
Infiltration barriers	
Terraces	Windbreaks/shelterbelts
Ponds	Catchments
Gabons	Anti-insect barriers

Waterbreaks	Corridors/habitats
Cajetes	Riparian buffers
Water channels	Firebreaks (living)
Mounds and beds	Living fences
Stone clusters	

Hedges and windbreaks are also examples of multi-threat counters. These can, to a lesser degree, retard evaporation when water is scarce. Well formulated windbreaks can offer slight protection during the extremes of temperate and, as positioned, they shield crops during high winds. Irrigation and windbreaks do not protect against all adversities and may be best when combined. With these anti-risk counters, more might be better.

Planting and Other Risk-Countering Strategies

Given anticipated changes in climate, some crops have a greater failure risk than others. The basic premise is to plant a variety of crop types. The hope is that all will yield. If climate or other factors intercede, the intent is that some may fail, but others, the most climatically resilient, will still produce. This means the planting of high, medium and low risk crops. Although not often formally recognized, this often involves a tiered strategy.

Another anti-risk planting strategies involve crop placement, i.e., field selection. Added tools include disarray, crop scattering, along with some site and/or crop-specific counters.

With these countermeasures, the basic idea is that, when weather is favorable, farming will be economically viable. When conditions worsen, the results may be closer to subsistence, but people and their livestock will survive.

Crop Tiers

At the center of a tiered strategy are the different crops. Foremost, and at the top tier, are the primary or staple crops. Rice, essential to billions of people, may be first on the list. The second most popular may be wheat. Although far less water demanding than rice, this crop requires more water that some other grains, e.g., sorghum or millet. Also high on the list of in-demand crops is maize. Even the drought-resistant varieties are, when compared to arid-zone crops, relatively water demanding.

As these can fail when rain is lacking, all fit within the classic tier approach. These are often upper tier crops, those most in demand and least likely to yield when growing conditions worsen.

Photo 10.1. A high-risk, high-tier crop supported through irrigation. In this case, wheat is being grown in an arid region of southern Africa that normally would not favor this crop.

A tiered strategy occurs when farmers seek to diversify risk by also raising a less climate-susceptible secondary crop. In the wetter parts of West Africa, paddy and/or upland rice is the upper tier, higher-risk crop. Cassava is a second-tier crop. It can still produce even if conditions are not right for rice yields.

There are other examples, in place of wheat, farmers may opt for sorghum or millet as middle tier crops.

In some regions, starvation occurs if both the primary and secondary crops fail. In this case, landusers may include, in addition to a middle tier, a bottom tier. The latter would be far-less susceptible to climatic variation.

The bottom tier crops are almost always productive trees or shrubs. Woody perennials are long living, grow with little effort, and, with deep, water-gathering roots, have less tendency to fail.

History is rife with examples. In ancient Europe, the acorn was the backup crop when weather, pillaging armies, or avaricious nobility took the staple grains. The more palatable chestnut, beechnut, and walnut would have supplemented a rather spartan acorn diet.

The tree species, moringa (*Moringa oleifera*), has been promoted in the Sahel of Africa. This has been done with a high degree of success. The notion being that the eatable leaves and seeds offer subsistence in the absence of staple crops.

There are other trees, from arid Africa and other regions, that grow in drier climates and will still yield when annual crops fail. The list includes

- argana (*Argania spinosa*)—A long-lived tree from southern Morocco that produces oil-rich nut.
- marula (*Scierocarya birrea*)—Found in southern Africa, this tree provides a fruit rich in vitamin C along with a nourishing nut.
- mongongo (*Schinzioplyton rautaneni*)—From southern Africa, this tree provides a nutritious nut.
- shea nut (*Vitellaria paradoxa*)—From arid west Africa, this also offers a fruit with nut.

As shown with the less than appetizing acorn, taste is not a requirement for good health. Nuts can provide calories, nutrients, and can be stored. The required desirable plant characteristics include low cost and low maintenance. Trees do this as, once planted, they require few, if any, inputs. In most parts of Africa, the only requirement is protection, when first established, from domestic or wild animals.

Those relying upon livestock may regard grasses as an upper tier crop. When rain is lacking, they may have second, less water demanding, forage crop planted amongst the moisture-loving grasses or interspersed in the farm landscape. As with most backup crops, this would be drought resistant tree or shrub. Species of atriplex (*Atriplex nummularia*) can insure grazing on dry, near-desert sites.

Tiers Quantitatively Expressed

Tiers are more a subsistence need and, as such, are quantitatively stated along these lines. The unit is calories or calories plus the vitamins and proteins required for good health. A rough equation is

$$K_T = (Y_a K_a)_1 + (Y_b K_b)_2 + (Y_c K_c)_3$$

where total harvested calories (K_T) equals the site yields from tiers 1 through 3, i.e., the site yield, e.g., from species a (Y_a). This is multiplied by the caloric value of a unit of this output, e.g., K_a. As conditions change, K_T should increase or, at the very least, come in at some minimum value. For simplicity, the above equation has single-crop tiers. The norm should be tiers with many productive species.

Photo 10.2. Recently harvested sorghum showing a mix of varieties. These come from multi-varietal plots.

Multi-Varietalism

Within the classic three-tier structure, variations abound. In drier part of West Africa, farmers intercrop different varieties of sorghum or soybean. The idea is to anticipate the uncertainty of a rainy season. In these cases, a short-duration, early-maturing variety is co-planted with long-duration, late maturing types. If the rains are briefer than normal, the early-maturing types will experience normal yields. If the rains are prolonged, the late-maturing variety will thrive.

This use of the varietal vector is risk-reducing measure as, if it were not employed, the farmers might miss the full impact of the rains and experience a worrisome shortfall in their staple crop.

Multi-varietalism is found in sub-Sahara with millet and soybeans. Again, the purpose is to hedge the start and/or length of the wet season and insure adequate yields.

The outcome is much the same as with other countermeasures. This results in a flattening of the yield curve (Figure 10.2). At the edges of the curve, where resources are in greater or lessor supply, mutil-varietalism can improve yields. In the optimal middle, the system does not do as well.

There is a drawback. When the produce is sold, the consumer might not appreciate visually-apparent diversity in their purchase. This is a perception, not a quality issue.

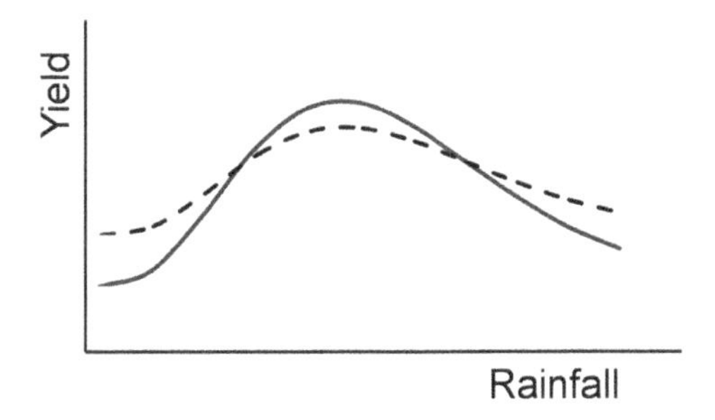

Figure 10.2. Production functions compared. The solid line is the normal, rainfall-related function without countermeasures. The dotted function shows what should happen when multi-varietalism, plowing irregularity, scattering, placement, and/or disarray are employed.

Disarray

Spatial disarray refers to the in-plot or landscape ordering of intercropped plants. Rather than a neatly ordered spatial system, i.e., plants in clear rows, blocks, or some other pattern, it is possible to place species in some disarrayed pattern. The effect is minor with a mono-crop, amplified through intercropping.

Disarray is integral with in complex agroecosystems (Chapter 11). It is also a risk reduction method.

Without the optimal density and spacial arrangement, yields will not be optimal. With disarray, some plants are closely spaced, others are at a greater distance. The result is that when resources are scarce, those plants with more space will compete less for water and will yield better. When resource are plentiful, those that are closer will offer greater per-area productivity. The net effect is to flatten the resource/yield production curve (Figure 10.2).

The general rule is that an ordered pattern is advantageous on sites where water, and other resources, vary within a narrow range. It is also advantageous where irrigation is available. Disarray is best where resources, mainly water, is a changing input.

Photo 10.3. Disarray in an Ethiopian household garden.

Plowing Irregularity

Another factor, hitherto researched, is uneven plowing. This is where many small puddles are present after a moderate rain. The effect, on the aggregate, allows for greater absorption of moisture. There is more.

Uneven plowing leaves some spots with high in-soil moisture, others drier. With crops, such as maize, a very wet season can favor those crops on the drier heights. A less well-water season will favor those plants in the wet troughs. The effect will follow the pattern as stated in Figure 10.2.

Placement

If in the right form, agrosystem placement can counter weather variations. This might include the uncertainties associated with climate change.

This involves not just putting water-demanding crops on wetter sites and drought-tolerant crops on drier sites. This approach, in not knowing what will happen in any given season, has water-demanding crops on both wet and dry sites. Crops that tolerate rain shortfalls are similarly situated.

This approach can be further reenforced by positioning some lower-risk crops on lower-risk sites.

In Europe during the middle ages, farmers had long, narrow plots. These plots did not often follow contours, but ran straight across rises and dales. This puts the one crop strip on a series of slightly varying sites. Cross-contours are illustrated in the lower left, Figure 12.1, p. 176.

One advantage is that it makes plowing with ox easier, i.e., there is less per-area, end-row turning. The primary advantage is that the long, narrow plots put each crop on slight different locations. With this cross-contour placement, some potential yields are sacrificed (as in Figure 10.2). Applied across many crops, both water demanding and drought tolerant, this has the effect of leveling yields.

The main disadvantage is a possible increase in the erosion threat. The other disadvantage is that these plots are not conducive for grazing.

Scattering

There is another aspect to placement, this is plot scattering. This has various manifestations.

Where population densities are very low, the landscape can have considerable non-cultivated areas. These can be where stretches of forests or grasslands separate crop plots. The classic example is slash-and-burn agriculture where, each season, farmers cut and burn plots within an intact forest. These scattered plots, if located in tall forest environment, are protected from temperate and, to a lessor degree, from a precipitation shortfall.

The scattered nature of the active plots means that there is the natural ecology from the surroundings (in the form of insect and disease control, water infiltration, etc.). Therefore, there is less incentive, and possibly less need, to provide surrounding defenses. This is magnified if the plots are continually relocated.

As population densities grow, the plots become less scattered, the between-plot, uncultivated areas smaller. This reduces the beneficial ecology.

Some scattering is due to topography. This is where the cultivatable land is limited and, out of necessity, plot must be scattered. This occurs in parts of Papua New Guinea where the mountain topography offers only small allotment of flat, tillable sites.

Scattering has other manifestations that are not limited by population density. This is where farms are not continuous, but each constitutes a series of widely-located, non-connected plots. This is best when the topography is diverse, usually with hills or mountains. This forces individuals to plant various crops in both high or low risk sites. Their attempt to reduce their own risk, combined with that inherent in the landscape, produces less risky agronomic landscape.

This was one element in as overall risk-reduction food strategy that was utilized by the Incas. Other elements included food storage and terraces.

Risk-Reduction Intercrops

There is a class of intercrops that have the ability to offer some, not all output when conditions worsen. This usually pairs a drought-resistant species with one from a higher tier. In above-average years, both crops yield. In the face of drought, only one species survives.

One reported example (Muuisse et al., 2012) is maize with watermelon. The melon provides the output in drier times. This combination would normally not find favor except as a low-rainfall counter. The two-output LER, in good years, being an unimpressive 1.13.

Fauna-Based Alternatives

Some may integrate animals in tier-based strategy. Villagers may plant grains before the rains arrive. If the rains do not come in sufficient quantity to allow the grain to set, the young plants become a secondary crop. These are directly grazed or the grain-less stalks can be harvested and dried as an off-season source of animal forage.

There are animal-centered, anti-risk strategies. Domestic herds, or flocks, can be increased when there is plenty of feed and forage. The animals are eaten or sold when grazing is difficult. As an overall scheme, this has drawbacks. Large herds can overgraze areas, causing environmental problems, e.g., erosion and the killing of trees. During bad times, when herd must be thinned, the selling prices can be low.

Nomads employ yet another anti-drought strategy. When grazing lacks, they simply move to another location.

Financial-Based Security

In most of the world, the economies are advanced and large enough so that starving people no longer rely on trees when crops fail. Instead, they rely upon purchasing power to have food shipped in.

This can comes from money saved in good times to purchase food when crops do not yield. Funds can also be borrowed. One problem, when crops are plentiful, prices are generally low, during food shortages, food prices are often high. Dispensing with the bottom-tier defense is unfortunate as it could and should be viable climate-adverse strategy, one that lessens the need financial strain on individuals and families.

Photo 10.4. The top picture is of traditional, grain-storage structures. The bottom shows a modern, government-funded, grain storehouse. Both photos are from the same village in Ghana.

Rather than saving or selling livestock, farmers can plant trees as a form of savings. This strategy is utilized where banks are not available and money must be saved for some distant event. Trees can also be cash reserve when crop yields are bad. This strategy requires nearby urban markets that continually seek firewood and/or construction materials.

Other Options

Peoples with access to nature-provided food may opt out of the tier-based food security system. Those in rural communities might have access to natural food sources and, in the short term, they can hunt, fish, and gather. For example, those along shores of seas or large lakes can go fishing and/or dig for shellfish. This option may be available to only a few.

A tried and true method is food storage. For this, staple crops, grasses and other needs are stored in bins or underground pits. There are traditional designs (as in Photo 10.4) that are adequate. Manufactured metal or concrete storage containers are more expensive, but can have fewer losses from birds, rodents, and/or insects. The disadvantage is the expensive.

Many societies, especially those that experience food shortages, adopt this as a partial or full solution. This goes pack to antiquity where central authorities encouraged grain storage, e.g., in ancient Egypt, China, and in the Americas with the Incas. This occurs to this day (also in Photo 10.4).

Tempting at this might be, storage is should not become the justification for high-risk crops. It is better not to confront the ultimate fate, but to take this away through a well-planned, anti-risk strategy.

Comprehensive Strategies

In high risk regions, it may be best to deploy a number of countering strategies. It is best to start with those that do not lower the yield curve, but shift it to the left (Figure 10.3). Drought-resistant crops, windbreaks, and various catchments accomplish this.

The other counter strategies flatten the curve (as in Figure 10.2). These measures include multi-varietalism, scattering, placement, and disarray. As always, this is an economic question. The best mix is determined by a far-from-refined assessment.

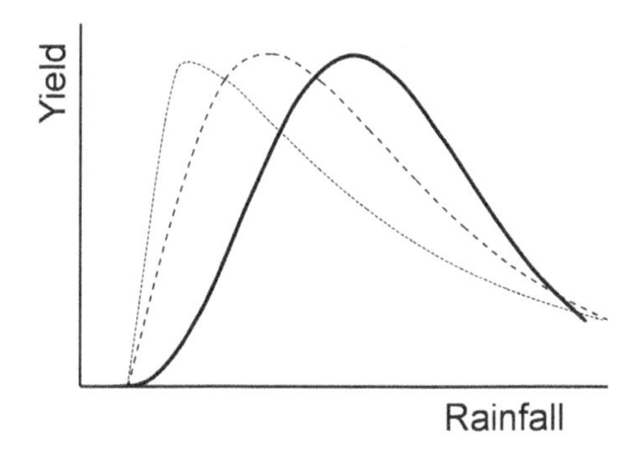

Figure 10.3. Production functions as these relate to risk. The solid line (right) is for maize and rainfall in the highlands of Kenya. The others (dotted lines) are theoretically projected to represent more risk-adverse crops.

Assessment

Looking at an already developed equation in monoculture form, R become R^K for comparing the riskiness of one crop, e.g., Y_a, against another. This is

$$Y_a = f(R^K).$$

For a single-variety monocrop where facilitation is ex-plot, e.g., from a distant windbreak or uphill infiltration ditches, the equation expands to

$$Y_a = f[(R + f(F_x))^K]$$

where F_x is external, cross-plot, facilitation. Decreasing the K-values shifts the function leftward. This is demonstrated with a different functional form.

Figure 10.3 (solid line) shows the relation between seasonal precipitation and maize yields in the Kenya highlands (Glover, 1957). A logical extension might suggest the middle, dotted line is a crop such as sorghum. A highly drought tolerant crop, such as moringa, could be the curve (also dotted) on the far left.

Employing the function suggested by Glover (1957), this is, based solely on rainfall,

$$Y_a = a\,(R^K)^b \exp(-c\,R^K).$$

With external facilitation added, $(R + f(F_x))^K$ replaces R^K as

$$Y_a = a\,((R + f(F_x))^K)^b \exp(-c\,(R + f(F_x))^K).$$

These equations are the basis for risk evaluation within the context of landscape design. These provide the numbers, extracted for linear programming, in functional form for non-linear problems. This could be an objective in multiple objective programming or a constraint with single-objective linear equations. As such, discussion continues in Chapter 12.

Summary

Through multiple crops and different types of livestock, tradition farms often incorporate, unnoticed, measures to counter weather-related, as well as economic, risk. The idea to not to be hurting when conditions are less than optimal.

With climate change on the horizon, risk avoidance can be, and in many regions, should be, part of a planned strategy. It may be best to utilize most, if not all, the countermeasures presented.

As a plan of action, there are obstacles. Risk is a topic where the supporting data is very scarce. Many of counters mentioned, e.g., placement, disarray, and scattering, have never been researched. Others are no more than mere mentions in the published literature. It will be some time before solid data appears.

11

Complex Agroecosystems

Chapter Preview

There is a group of agrotechnologies that stand apart because they are whole-ecosystem governed (right side, Figure 2.1). The opposite are those systems base on one-on-one, plant-on-plant dynamics. Ecosystem-governed systems are common in the humid tropics, these are managed through rule-of-thumb principles. With their superior environmental outlook, these have great unrealized potential in commercial applications, e.g., tree-crop plantations. A severe shortfall in research places the reliance on experience and the basic rules. To expand the possibilities, this chapter proposes theory in the form of two evaluation algorithms.

Chapter Contents

Introduction

The previous chapters have looked mainly at planned, managed, and ordered agrosystems. There exists a class of agroecosystems that exhibit high levels of biodiversity, but are also characterized by density and some degree of disarray. As a duplication of what nature does, to the uninitiated, these agro-systems can be mistaken for natural forests.

These come in varying forms and uses. Most common are mixed species pastures and managed natural forests. Both these produce one specialized output, either forage or wood. These systems are understood and techniques have been developed to maximize yields and profits and/or minimize costs. Because of this, they fall outside discussion in this chapter.

With the above exceptions, complex agro-systems are generally not suited for the large-scale production of one output. Most are at their best when they produce small volumes of many products. Outputs can vary and can be almost anything from annual vegetables to long-duration wood. This, along with their environmental compatibility, dictates their economics, their applications, and often their landscape location.

The homegarden version of the agroforest is common in the humid tropics. These are found near or surround dwellings. These have been the subject of some study where qualitative descriptions are commonplace (Nair and Kumar, 2006).

Another versions, the forest garden, exhibits the same complex structure, but is more commercially focused. These are found away from dwellings. These are also other, less encountered forms, e.g., shrub (as in Photo 1.2, p. 7) and forest gardens.

Although some descriptions of the different applications go into great detail, a key layer of quantitative data is missing. This is unfortunate as complex ecosystems may be an answer to some of the food and environmental problems facing mankind. The many hectares devoted to

Photo 11.1 (a and b). Two complex agroecosystems, the top photo is of a dwelling-surrounded tropical homegarden. The lower picture is of a forest garden. Both photos are from Bangladesh.

wood and treecrop plantations, e.g., oil palm and rubber trees, could be retooled as complex agroecosystems (Photo 9.1 shows such a situation). This would combine their intended purpose with additional biodiversity and additional outputs.

Because of the future potential, the agroforest form is of technical interest. In almost all aspects, from their potential, and unmeasured, productivity through to their environmental compatibility, these systems epitomize agroecology. As such, this is an area of agroecology that deserves further insight.

Types of Agrosystems

Given their complexity, eco-friendliness, and their low maintenance, this has lead to some distinct agrotechnological variations. As forms of complex agrotechnologies (described in Appendix 4), these are

- Natural pastures
- Mixed annuals
- Agroforests
  - Tropical homegardens
  - Shrub gardens
  - Forest gardens
- Enriched forests
  - Forest farming
  - Mixed cropping
- Managed natural forests

As mentioned, the first and the last are not under discussion in this chapter.

Economics

In agroecology, complex agroecosystems are quintessentially cost-oriented. These have very low levels of inputs coupled with what is assumed to be high and varied outputs.

Although fairly common in the humid tropics, they have defied data gathering and analysis. Productivity levels are not known and standard measures, such as an LER value, are missing.

This is due to the many outputs from many plant species. These are spread over a long time frame. The span is often multiple years. This is because the wood-producing trees do not have a yearly harvest. There is also difficulty in determining labor inputs. Much of the work done is casual, undertaken while gathering or harvesting an output, rather than as a dedicated, single-purpose task.

From a wider perspective, these systems can contribute much to a farm and a local economy. Nasser et al. (1994) reported that agroforests in Central America produce over 90% of the vegetable crop. The same importance is placed on the forest gardens in the South Pacific (Raynor, 1992).

Ecology

As stated, complex agroecosystems are ecosystem governed. This is in contrast to species-governed system presented in previous chapters. Ecosystem governance is where, except for minor, often unacknowledged labor inputs, nature takes over. There threats, including weeds, insects, and diseases, are suppressed through natural ecology. This reliance on natural dynamics is what gives these systems their high-level cost orientation. It also produces highly risk adverse systems.

These systems are a cornucopia of internal dynamics. Present would be all the mechanisms of competitive partitioning, including separate sources. Given the strong ecological currents, it would not be out of place to say that, given a free-form design, most, if not all, forms of inter-species facilitation also take place. The resulting ecosystem is greater that the sum of the parts. The parts being the total number of plant species.

As a duplicate of the natural forest ecosystem, these have all the inherent gains. All or, at the least, most of the mechanisms of facilitation are at play, insect and diseases are controlled mainly through an abundance of insect predators and a full ranger of micro-organisms. Nutrients are naturally recycled. High rainfall is absorbed and the ecosystem, as a functioning bio-unit, provides relief from drought conditions. Transpiration, shade, and other mechanisms also help mitigate extremes in temperature. If not overly small, these systems provide relief from the full range of productive threats.

In an era when climates are changing, there will be a premium on those agrosystems that are risk adverse. Agroforests are able to withstand conditions that would doom yields in, and the survival of, other systems.

There is more to the ecology than exceeding the sum of the plants. These also serve as a landscape refuge for bird and bat species along with other small fauna.

There are macro-gains. Global warming, deforestation, land degradation, loss of biodiversity, carbon sequestration, and food shortages can be addressed through tree-based complex agroecosystems. Their low levels of ex-plot inputs promote regional food security. This would be shown through a favorable output/input ratio.

As stated, visually and ecologically, these can be mistaken for forest fragments. Ecologically, these serve the same purpose within the larger farm landscape except that these are fully productive agronomic, rather than yield-limited or yield-inert natural ecosystems.

Rules of Management

As with intercropping, there are management rules or guidelines for complex agrosystems. As suggested by Wojtkowski (1993), these are

(1) if a plant output is needed and the plant is producing well, leave it; if not, improve the competitive environment;
(2) if it's production is not needed, neglect it;
(3) if it is negatively influencing a more desirable output, prune it; and
(4) if space exists and essential resources are under-utilized, as determined by the amount of light striking the bareground, plant or let something grow.

Under the above rules, seldom is a plant removed. Unwanted species are naturally suppressed, either through shade or a combination of shade and competition for belowground resources. Placed at a disadvantage, these plants, including invading weeds, eventually succumb.

The management rules for complex agroecosystem can produce good results. With an understanding gained from experience, seasoned users can take advantage of the inherent flexibility and achieve far more.

Basic Parameters

As with natural ecosystems, the complex internal ecology is some function of density (D_{den}), diversity (D_{div}), disarray (D_{dis}), and duration (D_{dur}). This can be abbreviated as D_4 such that

$$Y_{LER} = f(D_{den}; D_{div}; D_{dis}; D_{dur}) = f(D_4)$$

Density

The number of plants is a given area is a factor in agroecological design. In monocropping, there is a suggested number of plants per area. This can depend on the site, the crop and, variety. When two species are intercropped, both are not often spaced at their suggested monocultural densities. As demonstrated in Chapter 6, a lesser number of plants per area may provide the best yield possibilities.

This holds true for species complex designs and a yield density number can be computed for these polycultures. This density index (D_i) is, for a three species mix, computed as

$$D_i = (n_{abc}/n_a) + (n_{bac}/n_b) + (n_{cab}/n_c)$$

For this, n_{abc}, n_{bac}, and n_{cab} are planting densities for the three species, a, b, and c, when interplanted. The optimal planting densities for species a, b, and c, when planted as a monoculture are, respectively, n_a, n_b, and n_c. The rules of management suggest that, once all the direct light and some of the filtered and indirect light is utilized, a system is approaching the upper density limit.

Diversity

The number of different plant species in a given area constitutes diversity. This is also called agrobiodiversity or, simply, biodiversity. This refers to the number of productive and/or facilitative species with an agroecosystem. Unwanted additions, i.e., weeds, are included as these are facilitative in a functioning complex agrosystem.

In studied systems, the total species count ranges from about 6 to 200 species. The norm seems to be in the 15 to 30 range.

In ecology and agroecology, biodiversity is generally considered a good thing. Within the productive context of agroecology, diversity is a planned strategy, one expected to produce an economically positive outcome.

Diversity is commonly measured along two fronts. The first is the number of plant species present (richness), the second is the percent of each species in the total population of all plants (evenness).

It is the second, the percent of each, that provides system insight. Based upon the strength of percentage, three categories of biodiversity can be formulated. These are (1) primary species, (2) secondary species, and (3) trace species. Unintended weed species would (hopefully) be included in the latter category.

The primary species might constitute 25% to 75% of the total number of plants. These percentages would be roughly similar for the amount of biomass.

Part of this second category are those species that are fairly prominent, but do not have the population numbers and rise to the biomass levels of the primary species. Individual species in this category would be 5% to 25% of the total number of plants with similar percentages for biomass.

There are a large number of trace species where each species is represented by one or two individual plants. Each species would make up about 1 or 2% of the total species count. Although small by their individual populations and biomass, in total, the represent a significant percentage of the total number of species.

There is yet another factor, the percentage of tall species. If most of the light is captured by the taller species, the understory will be lost. This means a balance, or parity, between the tall, medium and short species.

There are a number of indices for formally measuring biodiversity. This can apply to both species richness and evenness (Coffey, 2002; Wikipedia, Biodiversity Index, 2013). At the core, all divide the number of species by the number of per area plants. With slight variation, these capture various nuances in the biodiversity. This may require an ecological, rather than an agroecological, indices, e.g., as proposed by Birrir et al. (2014).

Given all the diversity caveats, the presence of seven different plant species may be the point where species governance not longer sets the ecology and ecosystem governance takes over (Kareiva, 1994; Tilman et al., 1997; Baskin, 1994). To achieve all the competitive partitioning, separate source, and facilitative effects, number of species and their percent in the population should be at or above this threshold level.

The D_i is the total density of established agroforests. This is a key, but yet undetermined, number. For the full ecological functioning of agroforest, species evenness is helpful. Evenness is not in the number of each plant species, but is better expressed as the percent of each component species when compared with the optimal monocultural density for this species.

It may be better if these percentages equate as

$$(n_{abc}/n_a) \approx (n_{bac}/n_b) \approx (n_{cab}/n_c).$$

This is not strong statement, there exists considerable margin for flexibility.

Disarray

Spatial disarray refers to the in-plot or landscape ordering of intercropped plants. Rather than a neatly ordered spatial pattern, i.e., plants in clear rows, blocks, etc., it is possible to place species in some disarrayed pattern. In referring to planned and managed agrosystems, the terms biodisarray and agrobiodisarray are synonymous.

Photo 11.2. A dry land, disarrayed shrub garden featuring papaya, cassava, banana, and sorghum. The surrounding vegetation would be part of the active ecology. This photo is from Senegal. A higher rainfall version is shown in Photo 1.2.

In use, disarray should not be confused with randomness. Disarray is the absence, in part or in all, of fully patterned or fully ordered planting arrangement. Planning is still involved, either through the ground rules or through the knowledge possessed by the landuser.

Beyond a complete ordered, arranged system, disarray comes in varying forms. There are different degrees of partial disarray (with non-straight, but still discernible rows). There is full-on randomness.

Figure 11.1 shows, in overview, some of the spacial options. The upper right is of an ordered system. The lower upper left is semi-ordered, but still disarrayed. The lower left is of disarrayed block pattern and the lower right drawing is disarrayed with a mid-point design (further illustrated in Figure 11.2).

Other patterns are possible. Some would be variations of some base patterns, e.g., those illustrated in Figure 2.2, p. 17. Briefly, the patterns are row, strip, boundary, and block. Without getting too deep into the possibilities, there are examples where agroforests employ disarray with a base pattern. In Java, Jensen (1993) reported most of the plants crowded along the edges. The latter is possible if the individual gardens are small.

Compounding the complexity, there are two base canopy patterns. These are the midpoint and minimum interface design. These apply only to polycultures of three of more species.

The first of these has next tallest species placed midway between the tallest plants. This continues where each plant is at a mid-distance between the next tallest. This is illustrated in Figure 11.2 (right).

The minimum interface design has each plant in descending order based upon height, i.e., the next tallest is adjacent to the tallest, etc. The net effect is to have the taller is the center of a group where, going outward, each is smaller than the neighbor. This is shown in Figure 11.2 (left).

There is considerable margin to match species without disrupting disarray. For example, different species can be paired to take advantage of any inter-species complementarity. For harvest convenience, users can put like species in close proximity.

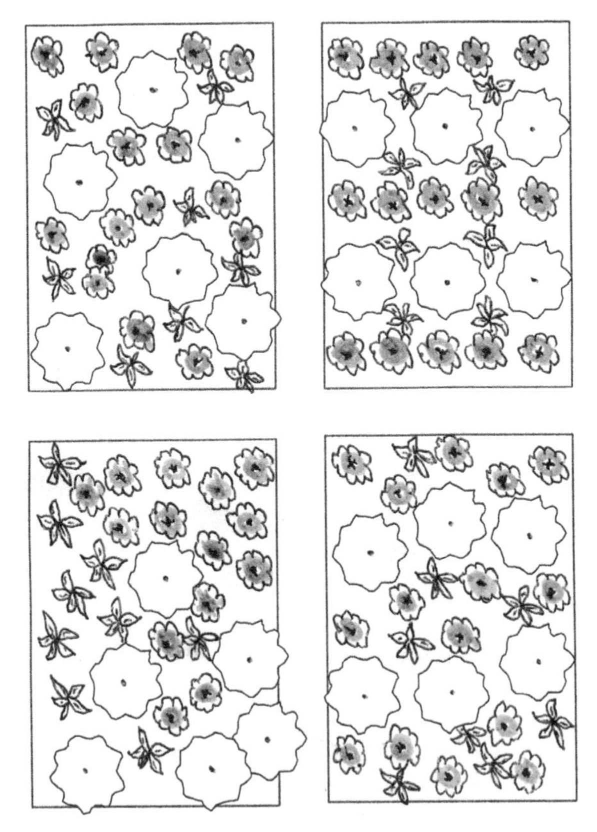

Figure 11.1. An overview of possible spatial patterns. There is an ordered (upper right), semi-ordered (lower right), disarrayed mid-point design (upper left), and a disarrayed block pattern (lower left).

Duration

The longevity of an agroecosystem, expressed as the time between major disruptions, can be an indication as to the proliferation of a range of organisms. Also referred to as agro-bioduration or bioduration, time allows a stable and viable ecosystem to develop, one that will hopefully be in balance and is contributory to a favorable farm outcome. This means an abundance of predatory insects hungry at the wait. It can also mean large populations of insect-eating birds and bats.

Duration allows more facilitative plants to enter. These are not always purposely introduced, but can naturally establish. Under the rules of management, these are allowed a brief stay. If they prove beneficial, they are kept, if not, they are eventually replaced with a plant species of greater worth.

Duration has another purpose. Given that these systems are biodiverse and disarrayed, there could, for any one plot, be hundreds of optimal solutions. Any user, with a modicum of experience, will, given the time, most likely find one or come close. In long-managed agroforests, one might assumed that, absent some extreme in the design (e.g., too much of one species or where like species are clumped) that optimization may have occurred.

Theoretical Development

It is not likely that the knowledge of an experienced landuser can be superseded by an algorithm. With small farms, anything beyond the user experience and the basic guidelines is not necessary.

In the study of advanced agroecology, this is far from a knowledge-for-the-sake-of-knowledge undertaking. Complex agrosystems have considerable potential. To become a commercial possibility, analytical progress must be made.

The LER Explained

Under agrosystem governance, when d_4 surpasses some threshold, the ecosystem become whole and the ecology is more than the sum of the parts. This relationship holds against all threats; insects, drought, high rainfall, etc.

In some ways, this invalidates the LER development in the previous chapters. If a complex agroecosystem was the sum of the parts, this would be expressed as

$$Y_{LER} = f\{[n_a (R - P_{b...m} + F_{b...m})] + ... + [n_m (R - P_{a...m-1} + F_{a...m-1})]\}$$

where m is the number of species in the system. There are practical considerations. For example, in tropical forests, the bulk of the nutrients are in the biomass, not in the soil. This makes it hard to set a nutrient-based R-value. This holds true with tropical agroforests. In addition, the complexity of these systems make it difficult to apportion light and water.

Ecosystem dynamics do not totally invalidate the expanded-LER equation. This still has use as an explanatory tool. It is useful to examine the dichotomy of thought. As developed here,

$$Y_{LER} = f(D_4)$$

and, with the expanded-LER version,

$$Y_{LER} \geq f\{[n_a (R - P_{b...m} + F_{b...m})] + ... + [n_m (R - P_{a...m-1} + F_{a...m-1})]\}$$

There is a relationship between the two equations. It is assumed that ecosystem governance occurs when plants are in close contact (density). Ecosystem governance, competitive partitioning, and facilitation cannot come about unless the number of species (diversity) exceeds some threshold.

Also, ecosystem governance, competitive partitioning, and facilitation cannot occur unless there is a high degree of interaction between unlike species. Disarray insures that this happens. Duration carries the assumption that time is need to maximize the sum total of the competitive partitioning and facilitation effects.

Density, diversity, disarray, and duration assume that competitive partitioning and facilitation are in full force. Likewise, competitive partitioning and facilitation assume threshold levels of density, diversity, disarray, and duration.

Reconciled, the LER equation is

$$Y_{LER} \geq f(D_4)(f\{[n_a (R - P_{b...m} + F_{b...m})] + ... + [n_m (R - P_{a...m-1} + F_{a...m-1})]\})$$

For this, the D_4 values are assumed be at 100% or, in ratio form, unity. These then fall out of the above equation. Lesser values, those below the threshold, are greater than zero and less than one. These would be undetermined, nonlinear equations.

If this line of development were to continue, disarray and the differing interactions associated with like species forces Monte Carlo simulation. This would be a m^2 endeavor with random values defining the relationships between the individual each plants. This is rather cumbersome, more suited to intercropping than highly complex systems. There is a more insightful, less involved algorithm.

Apportioning Essential Resources

With an agroforest ecosystem, some species are favored, some are relegated to lesser role. This is done through both location and the allotment of resources.

On small farms, sunlight apportionment visually guides the system. In accordance with the user guidelines, the management options are mostly pruning and thinning. Under these informal, visually-guided pruning regimes, any disarrayed pattern is possible.

At the commercial level, managers require a more inclusive, less hands-on forms of management. This is where canopy patterns will come into the fore. There are two core patterns, midpoint and minimum interface (Figure 11.2).

The minimum interface (lower drawing) directs more of the essential resources to shorter, understory plants. This happens because a bio-diverse understory, one that receives sufficient light, can compete favorably against the taller trees for belowground resources (Malcolm, 1994). This assumes that the lower-level plants are crowded against and encroach on the entire root zone of the taller species.

In contrast, the midpoint layout would, with a high density, narrow spacing, favor upperstory trees as less light reaches lower levels. With more open spacing, the understory has the opportunity for a better yield balance.

There are combination patterns. Commonly found is a midpoint design for the uppermost, taller plants. Those nearer ground level assume a minimum interface pattern. The distance between plants is part of apportioning light and other essential resources with the agrosystems.

The Allocation Algorithm

Complex agroecosystems can be managed by an equation-defined relationship. It holds that, for each individual plant,

$$(1/p_a)\, Y_a' = (1/p_b)\, Y_b' = \ldots = (1/p_m)\, Y_m'$$

where p_a through p_m are the ratios of the selling prices or relative value of the outputs. Y_a' through Y_m' are the derivatives of the production functions, commonly for light, for each species. This is similar to the maximization of inputs (Chapter 5) except that the full, not the decision range, of a production function is within consideration.

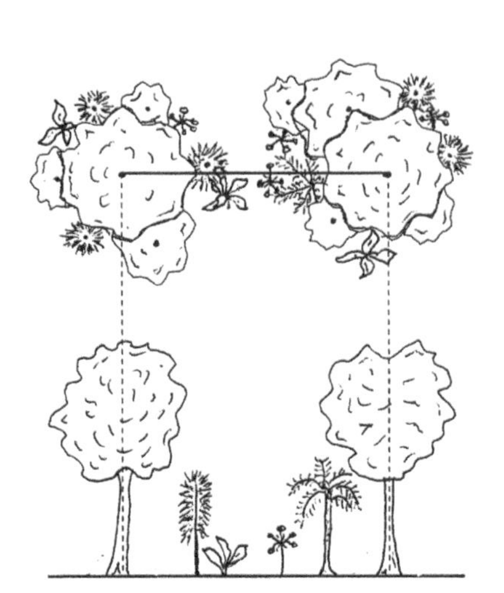

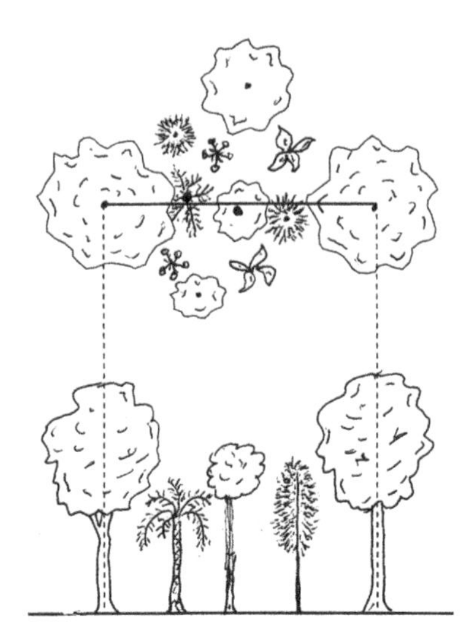

Figure 11.2. The two core three-plus canopy patterns. The right drawing shows midpoint design in overview (top) and an along the line profile view (bottom). The left drawing is a minimum interface layout. Again, an overview (top) is shown along with line-defined profile of the design. These might be the basis for commercial application of agroforest form.

As an example, the Allometric function $y = L^{0.5}$ has the derivative $Y' = 0.5L^{-0.5}$. From this, the economic allocation of light (L) for two of the species in an agroforest setting, would be

$$(1/p_a)(0.5L_a^{-0.5}) = (1/p_b)(0.5L_b^{-0.5})$$

It is more convenient to proportion the p-values with a range that equates with, but is the inverse of, the slope values.

Assuming that species a is far more valuable than species b, then, for these to equate, species a would receive far more resources that species b. It should be noted that, if the equation is a sigmoidal function, the slopes equate on each side of the inflection point. Using the second derivative avoids any confusion.

It is important to note that, in the sigmoidal form, complex agro-systems operate below the inflection point. This may seem antagonistic to good economics (as with the decision range, Chapter 5). Remembering that these are cost-oriented systems, the cost savings from the D_4 parameters, specifically density, mean that per-plant output is at the lower portion each plants production function. Moving to the upper reaches of this curve will greatly increase costs and change the ecological profile of the complex agroecosystem.

The Sustainability Constraint

For any complex system to yield into perpetuity, the nutrient (N, P, and K) outflows must not exceed inflows. This is a limiting factor on the overall agroforest and on the individual plants. The nutrient outflows, system exported from each plant, would sum at or below a set number, i.e.,

$$(K_a)(Y_a) + (K_b)(Y_b) + \ldots + (K_m)(Y_m) \leqq \text{sustainability limit}$$

The K-values represent the per unit output of potassium for each species. Parallel functions can add P and N. Y_a through Y_n are the total units (yields) from each species.

In practice, users have ways to increase system exports without losing sustainability. For homegardens located near or surrounding households, people tend to dump their organic waste in the agroforest ecosystem.

Beyond the adding of external inputs, there are two techniques to increase the active and exportable yield. One is the nutrient content of the output, the second is the non-harvest option.

All outputs take nutrients (N, P, K, etc.) from the system. This can involve planting more of those species whose yields have a low nutrient content. On the aggregate, this allows greater system-exported productivity.

Fruits and vegetables tend to be lower in nutrients, per kilogram, than staple crops. Comparing values, the percent of N, P, and K for oranges is 0.12%, 0.08%, 0.48%, for cherries, it is 0.18%, 0.06%, 0.2%, and, for potato, 0.21%, 0.07%, 0.29%. Wood is best. Without the bark, it takes almost nothing.

In contrast, staple crops, such as rice (1.08%, 0.18%, 0.09%) and wheat (1.93%, 0.87%, 0.55%), generally remove more (the nutrient values are from Roberts, 1907).

The second method is the non-harvest option. This where the vegetables, nuts, fruits, etc. that are not needed, are of poor quality, or are of low market value, are not harvested. Instead, they are allowed to fall and decay.

The effect is to recycle the nutrients. The notion is that these same nutrients will be harvested, again in a different, more valued and more useful form. This method is commonly encountered. The non-harvest option, added to the already complex system, makes it harder to determinate individual and overall yields.

Another Approach

For all ecologically-realized agroforests, the D_4 parameters must be met. Beyond this, the allocation of essential resources, along with D_i and sustainability, form the core constraints. Along with an objective function, content and density are resolved through non-linear mathematical programming.

Objective Function

This objective function for this system of equations reads as

$$\text{objective function} = \max (p_a\, n_a\, Y_a + p_b\, n_b\, Y_b + \ldots + p_m\, n_m\, Y_m)$$

For this, the number plants in each species group are n_a through n_m. Instead of seeking to maximize species populations or yields, a mix of both is sought. This is a relative ranking that, within the confines of a complex agroecosystem, seeks to rank based on per-area yields for the different species.

For example, if the component species are ranked alphabetically, by importance or economic value, i.e., a being the highest valued, m the lowest, then the relative ranking is

$$(n_a\, Y_a) > (n_b\, Y_b) > \ldots > (n_m\, Y_m)$$

For the most favored, highest-valued species (no more than two, possibly three), the plan is to put more resources (generally sunlight) to great number of the valued plants. These are the primary species.

For species of intermediate value, each of the species can be fairly large in number, but with few allocated resources. Each species can also be small in number with greater access to resources. These are the secondary plants.

Generally, there are a lot of lower-valued species. Each is few in number and must compete for available resources. These are the trace species.

This is a judgmental process. If the objective ranking for one species is very high in comparison to others, the population of the one species, as a percent, would be in the 30%–40% range. As the preferred species, these plants would be positioned to receive more light or the neighboring plants pruned to allow this to happen.

If the ranking to low in comparison to others, the plant would be neglected or pruned back. This follows the stated rules of management.

Constraints

As a mathematical programming (optimization) problem, the objective function is subject to constraints. These are

$(1/p_a)\, Y_a' = (1/p_b)\, Y_b' = \ldots = (1/p_m)\, Y_m'$
$(K_a)(Y_a) + (K_b)(Y_b) + \ldots + (K_m)(Y_m) \leqq$ sustainability limit
$L_a + L_b + \ldots + L_m \simeq 100\%$,
$n_a \geq$ lower limit (number of plants),
$n_b \geq$ lower limit (number of plants),
etc.

For the constraints, the Y′-values (Y_a' through Y_m') are the first derivatives of the light functions, i.e., $f'(L_a)$ to $f'(L_m)$. This is appropriate as agrosystems are often visually managed primarily by apportioning light to each plant.

At the upper end of the sigmoidal range, Y_a' through Y_m' should not be too small, this indicates marginal inefficiencies. There are also direct inefficiencies. In accordance to the resource-use hypotheses, e.g., von Liebig, Mitscherlich, etc., a shortfall in another resource means that a full allocation of light is not utilized with meaningful gain.

The lower points on the curve represent a shortfall of essential resources which, if too small, may lead to the eventual demise of the species in question. Under the rules of management, the lowest-valued plants are mostly permitted, but seldom removed.

The amount of a resource used, in this case light, should not exceed some maximum. With all resources, the total, due the biodiversity gains and ecosystem efficiencies, can exceed 100%. For example, light rays pass through otherwise dense foliage or are reflected off leaves onto the leaves of other species. These ecosystem efficiency gains puts the effective sunlight value at slightly over 100%. Such gains also occur with water and the belowground essential resources.

Lower limits on species biodiversity are in place (e.g., $n_a \geq$ lower limit). This is simpler than employing an evenness index.

Results

With regard to agroforest-type systems, there is a dilemma. Quantitative field data is hard to gather and, as a result, none has been forthcoming. To entice greater utilization, commercial applications require more than the rules of management. Large scale field trials are less likely, hence, pre-use analysis can only be accomplished through algorithms and predictive models.

For many candidate species, experience fills in many of the parameters for light-based functions. This is only a small step. The main issue is that the key system-diagnostic or system-deterministic values are missing. These are the density index, sustainability limits, and the expected LER. Without calibration, simulation models are predicatively adrift.

Even with the unknowns, there is some value in this analysis. The decision variables are the allocation of light and positioning and populations of each species. From this, the algorithm can suggest a layout and a management plan. There is considerable flexibility built into the system. If the decision variables prove slightly off the mark, a change in the management plan would allow, over time, these to be site adjusted.

Rules/Management Gap

Across the span of agroecology, the amount of biodiversity, and therefore agroecology, vary. In rough order, these are

- monocultures
  - seasonal
  - tree-based
- intercropping
  - seasonal
  - tree-based
- agroforests

The tree-based systems, as perceived mono- or intercrops, can contain greater levels of diversity. This often comes as a covercrop or as a tolerated, weedy understory.

There are rules of management for intercropping (Chapter 6). These rules are best with two of three productive species, but can be stretched to include up to four or so species. These can be crops or trees, not both.

At the other end of the spectrum are the rules for complex agrosystems (this chapter). These require at least 10 productive species for effectuation.

There is a large, unoccupied conceptional/management blank going from the intercrop to the highly complex agroforest. This is the agrosystem management gap.

For commercial endeavors, the methodology presented in this chapter will help to fill this void. This can be with 4 or 5-plus polycultures, i.e., those of four, five, or more species. These can be either disarrayed or ordered.

A semi-ordered pattern, as in Figure 11.3, may be best for the commercial use as suggested below.

Figure 11.3. A possible commercial spatial design, in cross-section, that would allocate light and other resources to different species in relation to their relative value. Those marked with arrows are the primary species. Between these pairs of palms are harvest alleys.

Spatial Layouts

Layouts should be based on the amount of light that should be allocated to each plant. Figure 11.3 shows how this can be done with a primary species (four plants) and other, secondary species. Figure 11.3 is a mix of mid-point and minimum interface designs. This helps allocate resources according to a plants relative worth.

Rule-Based or Spatially-Based Management

The methodology proposed here can guide all complex agrosystems. There is still the practical question of how to actually internally manage these systems. Rule-based management assumes a relatively high level of labor inputs. It should be noted that per area, and even per unit of output, labor is still low in comparison to intercropping. This, coupled with the need to train field workers, makes these less attractive beyond a small farm setting.

The other option is spatial management. The two base patterns, as in Figure 11.2, might help. As briefly mentioned, a combination pattern is possible, e.g., midpoint for taller plants, minimum interface for the those in the understory. If commercial applications are to succeed, these patterns must be understood and, from this, utilized to their fullest degree.

Potential

In many newly-opened agronomic landscapes, there is a problem of forest fragmentation. As a continuation of the same topic, commercial tree or treecrop plantations are criticized for rendering large areas inhospitable to local flora and fauna.

There is opportunity to do more. Across these large spans, the single-tree agrosystem can be replaced by a series of agrotechnologies. Tree or treecrop monocultures transition to tree or treecrop-based, five-plus polycultures which, in turn, transition into agroforests. The latter have the tree or treecrop as the primary species. Adjacent these are the forest fragments.

Another option is to dispense with the tree or treecrop monoculture, instead, to rely entirely on a mix of five-plus polycultures and full-diversity agroforests. Again, these have the tree or treecrop as the primary species. As mentioned, this could provide a wood or treecrop output along with a mix of foodstuffs. The spacial layout and the management would favor the primary crop.

When plantation managers do not want added work beyond the primary species, there are opportunities for multiple participants. With the advanced analysis demonstrated here, gains can be equability allocated to both groups. Equally important, the resulting landscape would be an environmental improvement over the single-species monoculture.

As mentioned in Chapter 9, industry relies on a set in-flow of raw materials. Any changes to these mono-plantations must show that added complexity need not interfere with, and might augment, the availability of raw materials.

Summary

Given the lack of data, this chapter is somewhat speculative. Even a modicum of information, especially a range of system LER values, would go a long way to redirect or solidify that presented here.

Despite their popularity, complex land-use approaches are mostly unrealized ideals. The agroforest form, being the ecological epitome of agroecology, should not go underutilized. The greatest potential comes if expanded into the commercial sphere.

12

Landscape Agroecology

Chapter Preview

As one of the many dimensions, a well formulated landscape, rich in ecology, can augment or substitute for good plot ecology. These gains should not be overlooked. Presented are rules of landscape design that aid the process. More formal analysis would also cement the often overlooked benefits.

Chapter Contents

Introduction

Agroecology extends beyond the individual plot to the multi-plot or landscape level. As broader economic and ecological platform, the landscape is not a brief, nor an easy topic. Many manifestations come into play.

Ideally, an agronomic landscape, if well formulated, can be more than the sum of the individual plots. This is due to shared and favorable, cross-plot agro-dynamics. A prominent example is insect-eating birds and their nearby habitat needs. Bio-complexity underwrites these inter-plot dynamics.

This leads to the notion of landscape solution theory. There are many ways to subdivide and plant an agronomic landscape. Correspondingly, there are many optimal solutions for a landscape layout. It may not be possible to identify all these solutions, but understanding a few, and how these are arrived at, would be a great help.

Objectives

As with plots, landscape have socioeconomic objectives. These are often more broadly defined than those for individual plots. Restated from Chapter 3, the principle objectives are

profitability (revenue-costs)
risk reduction
economic orientation
resource (labor and expense) allocation
return on investment
environmental

As is often the case, the landscape must return a financial profit. For subsistence farmers less concerned with the balance sheet, the farm must produce a decent living and each plot is so gauged.

In pursuit of favorable landscape ecology, a single objective, e.g., profitability alone, is not sufficient. Some of the criteria that underlie single or multiple-objective analysis include, besides profitability, economic orientation, agroecological-fostering biodiversity, and environmental concerns.

Multi-Plot Analysis

In Chapter 2, a series of ratios are presented. These are, in addition to the LER, the RVT, CER, adjusted CER, and the EOR. These diagnoses can be for a single plot or can apply across a many-plot landscape.

Often interesting is the economic orientation of an entire farm. This is a plot-weighted sum where the Landscape Economic Orientation (LEOR) is, for plots a_1 and a_2,

$$LEOR = A_{a1}(CER - RVT)_{a1} + A_{a2}(CER - RVT)_{a2}$$

where A_{a1} and A_{a2} are the areas of the two plots. The total area has $A_{a1} + A_{a2} = 1$.

Similarly, the cost efficiency of the overall landscape can be evaluated by way of the adjusted CER, i.e., CER(RVT). Expanded into a two-plot equation (LCER),

$$LCER = A_{a1}(CER(RVT))_{a1} + A_{a2}(CER(RVT))_{a2}$$

When each of two adjoining plots have 50% of the area and the CERs and RVTs are set to 1.0, this produces the following

$$LCER = 0.5((1.0)(1.0))_{a1} + 0.5((1.0)(1.0))_{a2} = 1.00$$

Redesigning such that the two plots are combined. To this now larger area, a series of strips are inserted. The purpose may be to reduce a less costly form of insect control. Usurping 10% of the area for these strips, the resulting overall cost is assumed to decrease by ¼. The CER(RVT), i.e., 1/0.75, becomes 1.33. The LCER is

$$LCER = 0.9((1.33)(1.00))_{a1} + 0.1(0)_{a2} = 1.20$$

This example shows 20% cost-efficiency gain through this change. This equation would also have application with other ratios.

Landscape-Specific Agrotechnologies

All agrosystems, whether productive or purely facilitative, influence the overall ecology of an agronomic landscape. The facilitative agrotechnologies are specifically tasked to provide a benefit. Some are plot internal, e.g., mounds and micro-catchments. Most function by providing benefit to

Photo 12.1. An in-plot infiltration ditch designed to increase in-soil moisture and extend a cropping season further into a dry season. The intent would be higher yield in this moisture constrained region. This photo is from Chile.

neighbor or nearby plot. The purely facilitative landscape agrotechnologies are listed below. All are described, in detail, in Appendix 4.

infiltration barriers	stone clusters
terraces	stone fences
paddies	windbreaks/shelterbelts
ponds	catchments
gabons	anti-insect barriers
waterbreaks	corridors/habitats
cajetes	riparian buffers
water channels	firebreaks
mounds and beds	living fences

Solution Theory

Agronomic landscapes are at their ecological best when they contain large amounts of biodiversity. It is possible to have a free-form landscape where the biodiversity is scatted without the use of formal agrotechnologies. In the majority of cases, landscapes are best as a series of varying agrotechnologies.

It is better if each agrosystem adds to the overall amount and to the ecological impact of biodiversity. This involves determining how much agroecology to incorporate and which agrotechnologies are most welcome in this regard.

This process is not entirely open and the objectives not the only user criteria. There is a host of mitigating factors. Many of these keep landusers/landowners on set tracks and away from unimpeded optimization.

Limitations

In contrast to individual plots, large and even small-farm landscapes are seldom empty expanses to be arranged and filled at a landusers discretion. There are many external and internal land-demarcated requirements that must be addressed.

On the list are the types of crops raised, the economic orientation, local divisional patterns (the positioning of legal and natural boundaries), land control (imposed, informal, or agreed upon ownership or rental constraints), topography, climate. Also on the list are gender considerations and cultural values.

It is not unusual for a farm landscape to be topographically circumscribed such that the landuser does not have full control over the internal ecology. For example, ownership boundaries can horizontally transect hillsides. If inter-farm cooperation is lacking, this makes erosion control efforts difficult.

Also, in regions of small farm plots, the application of a toxic insecticide in one area can destroy the favorable natural insect dynamics for all. Being rooted in nature, time, tradition, and through prevailing economics, farmed landscapes, once set in place, can be hard to change.

Investment

In agroecology, investments come in varying forms. One soil input, biochar (in-soil charcoal), is a permanent asset that gives a return as long as crops or treecrops occupy the site. With less gain from enhanced soil properties, the return on this investment is lower if forest trees are planted. There is no return if the land goes unfarmed.

Most of the land modification and bio-structures can be investments as long as agronomic activity continues. An example is the use of terraces. These help trap water, prevent erosion, and make cultivation easier. With the right crops, terraces provide years, even centuries, of advantage. Without yearly cropping activity, the economic benefits are greatly diminished.

Variations of the basic agroforest, once established, are costly to get rid of. Part of this is stump removal. This tends to encourage a tree-filled future. There is latitude for change without entirely loosing the return on this investment. By slowly altering the management and the species content, the investment is not lost, but the output mix is redirected to reflect new economic realities.

This is also the case with tree or treecrop plantations. There may be no need for an immediate, traumatic halt. A multi-stage taungya approach allows for a gradual shift in species and output. This continues the return on the original investment.

Factors of Likeness

The likeness between farms is the result of a concordance of forces. The factors of likeness are

- investments (in land-modification and bio-structures as previously described, also including crop-specific farm equipment and farm structures),
- governmental (extension agents and/or government policies or payments that push one crop and/or one agronomic form),
- societal (where the community exerts a subtle influence that results in everyone planting the same crops in similar ways),
- commercial (e.g., a thriving trade in one high-value commodity),
- user knowledge (where there is limited cognition as to the agronomic possibilities which causes all to raise only one or a few crops), and/or
- physical presence (the soils, topography, and climate that favor or force one agronomic expression).

In contrast, less uniformity might indicate a society in flux, markets that no longer exist, or crops that can no longer be grown. There are examples. An immediate change occurred in Costa Rica when the abandonment of a railroad resulted in the loss of markets (Soluri, 2001). More common is where aquifers are pumped dry and irrigation water is no longer available.

Rural Population Density

Landscape design is very much a function of population density. As this increases, the individual farms take on a likeness. This is because all or most of the area is needed for commercial crops and the necessities of life. There is little room for experimentation or individual expression. Biodiversity, both in number of species and agrotechnologies, is maintained long as this is perceived as the most efficient means to produce the needed crops. The need for greater productivity can favor revenue-oriented intercrops.

At the other end of the spectrum are those areas with far fewer people in a given region. These are more of a blank slate upon which to write a landscape design. Although limited commercial opportunities may direct landusers to one crop type, biodiversity expressions can still manifest. There can be a greater emphasis toward cost-oriented agrotechnologies.

The exceptions are those areas taken over by very large industrial (factory) farms and commercial plantations. Most often, these are the highly revenue-oriented, single-crop systems that offer few, if any, expressions of biodiversity. There is large potential for agroforest-type systems. With no prototypes to follow, this will be difficult to realize.

Agroecology expression can be greatest in urban areas where markets will accept, and demand, a wide variability in farm outputs. It is especially helpful if there are a mix of cultures, culture-based diets, and a variety of foods are consumed.

Design

The layout of crop plots and the resulting inter-plot ecological interactions can be a forceful tool in achieving ecological harmony. This covers the entire gamut of threats.

Herbivore insects do best with large expanses of a single crop. They do less well if an area is biodiversity transected and they do not have unimpeded access.

Along these same lines, with good habitat possibilities, predator insects can thrive. These dynamics are at their highest when the plots can be quickly (re-)stocked with the predator types. Some of this is accomplished with smaller or long, narrow plots interlaced with natural strips and corridors. These allows insects to bred, move freely, and range widely.

These same dynamics address other threats, lacking a large area of a single crop, plant diseases find it more difficult to spread. With a well designed landscape, these gains occur at little or no cost.

The anti-flood and anti-drought measures are also best landscape wide. Strips and corridors, if well positioned, can be part of water infiltration plan and/or part of anti-erosion strategy. Despite some land being lost to direct cultivation, this may not results in a decline in overall yields and profits. Small, plot-internal, bio-structures provide enough in the way of ecological services to overcome any potential loses.

An ecological balance offers other benefits. The preservation of local flora and fauna is best undertaken by managing the full area. Included in an agro-landscape are perennial sections where the local plants reside and insects, birds, and other native fauna roam.

Favorable ecology originates in and flows from bio-rich productive plots and/or from non-productive landscape bio-structures. With this in mind, the best farm landscapes may be where ecologically-weak plots neighbor and are fortified by way of the ecological spillover from the ecologically strong plots.

Facilitative plots or auxiliary systems, although not productive in their own right, benefit neighboring plots. Either through revenue increases and/or cost reduction, the gain in profit should be apparent. This comes about by maximizing the interface between the two plot types.

This is not all ecological, landscapes can also intensify the economic gains. With crop diversity, farm labor is more efficiently utilized if spread out over time rather than being concentrated in one short, single-crop planting and/or one short harvest season. Farm machinery can be smaller, and less expensive, if there is less pressure to harvest large areas in a short time period.

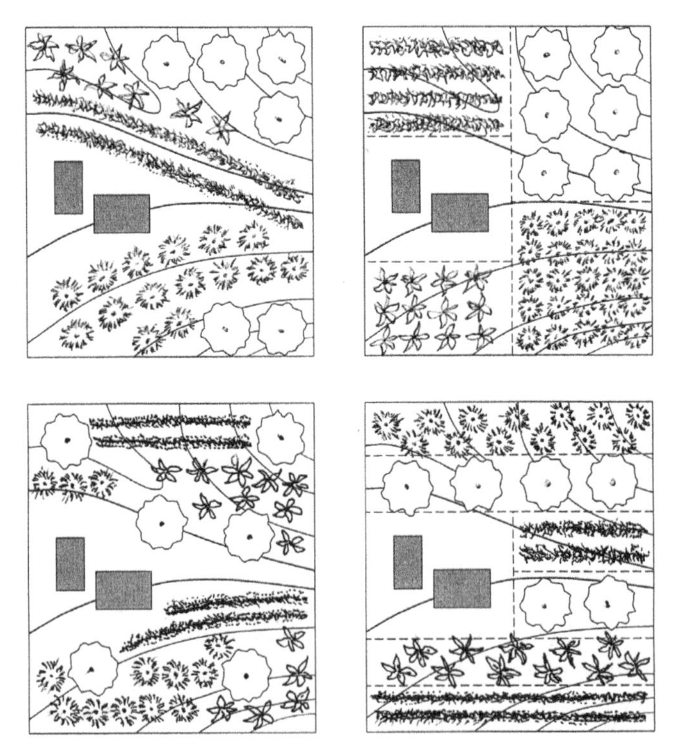

Figure 12.1. Different landscape layouts, contour based (upper left), geometric square or near square (upper right), free form (lower left), and strip (lower right).

Landscapes Types

Capturing the full inter-plot ecology requires a suitable overall landscape design. The vast majority of existing farm landscapes are geometrically expressed with near-square plots (Figure 12.1, upper right). Dispensing with the square or near-square shapes, also possible are strip patterns (lower right, Figure 12.1).

A more free-form, non-geometric, contour-based layout (upper left, Figure 12.1) follows the land contours and, at the same time, increases the amount of inter-plot interface. Hopefully, this will result in more inter-species ecology.

Inter-species ecology can be maximized by emphasizing disarray. This is done with an entirely free-form layout (lower left, Figure 12.1). In this figure, each section contains the same relative amount of each species represented.

There is no one best pattern, each has purpose. The strip and the disarray patterns are slightly better at countering risk. The contour pattern helps alleviate erosion danger. The more-common, near-square design may have gains for grazed systems or in less costly harvests. Broadly, this is a topic in need of further study.

The rules of landscape design (below) are accepting as to the landscape type. Generally, they lean more toward the contour or strip patterns.

Rules of Landscape Design

It is possible to carry forward the design concepts and formulate rules/guidelines. The overall goal is mitigate threats and to fully exploit landscape-wide ecological activity. These rules, below, are ordered by their relative importance. As proposed here, these are,

(1) on non-level, erosion-prone sites, rows are oriented parallel to the hillside slope,
(2) if there are no erosion/infiltration issues associated with a proposed agrotechnology, e.g., complete ground cover, or if located on flat site, and
    (a) if there is a large height differential between the component species (by a factor of two of more), rows should either be oriented north/south or to maximize early morning light (northeast/southwest in the northern hemisphere, southwest/northeast in the southern hemisphere),
    (b) If there is no significant intra-species height differential, plot/row orientation is set based on plowing, planting, and/or harvesting needs,

(3) so that interface dynamics penetrate across, and affect, most of the cropping area, the direction of orientation (the plot length) is proportionally longer than the non-orientation direction (the width), except when
  (a) there is no between-plot barriers to protect against profoundly negative interface dynamics,
  (b) active grazing is on-going or part of the rotational sequence, or
  (c) forest trees are the primary species,

(4) to maximize spillover, plots are positioned such that the ecologically-dynamic plots share a long-as-possible border with those that are less ecologically active and in need of an (agro)ecological contribution.

These rules can be ecological as well as economic. For crop plots, the length can greatly exceed the width making for long and narrow plots. This can reduce plowing costs as tractors have longer runs with less, per-area turning. Size and shape must be balanced against the pro's and con's of the harvest and the gains from more inter-plot ecology.

Animals graze more efficiently and per area fencing costs are less with square or near-square plots. For most forest trees, side branching is greater along the open, sunlit edges. This can reduce the value of trees grown for lumber. This is less important when firewood is the goal. In addition, tree branches tend to overhang and shade bordering plots. Without a branch-preventative interface, e.g., a specific edge or boundary species or a bordering treecrop, edge branching is normally reduced by minimizing the amount of interface with open crop plots.

Exceptions

The above rules assume a plot-based landscape. These are areas of a single agrotechnology, either as a monocrop or a more biodiverse expression. A few cultures forgo the plots and formal agrotechnologies, instead they adopt free-form, plant-based layouts.

For this, the rules are simpler. These are,

(1) plants or groups of like plants are located where they grow best provided this does not eclipse more productive land uses, and
(2) the amount of a plant species in the landscape is proportional to its value, except where constrained by soil and risk factors (Wojtkowski, 2004).

Free-form landscapes are more the exception and are mostly found in parts of Africa. These forms are not exclusive. It is possible to combine plot-based and free-form designs.

Photo 12.2. A hillside garden plot demonstrating contour planting. In this high rainfall regions of central America, the purpose is to avoid erosion.

Landscape Optimization

There is a rather elementary exercise where, given a set of constraints, the landscape is area optimized around an objective function. In accomplishing this limited analysis, profit is often the single objective. The constraints relate to farm topography, soil, plot sizes, etc.

Without any acknowledgment that the farm is more than a profit center, the result is intuitive. The suggested solution has a farm entirely covered by the single, most-profit landuse.

Many farmers and other landusers are generally more environmentally enlightened. Most realize that there is more to farming than a single objective and a single agrotechnology. It should be noted that factory farm owners are an exception.

In modeling complex farm landscapes, multiple optimal or multiple near-optimal solutions will be the norm, not the exception. It is also quite probable that these solutions will vary greatly in the range and application, i.e., the design of one farm solution could vary greatly from the next, even though those the farms are of similar in size, layout, in objective needs, and physical presence (e.g., topography and/or soil types).

The optimization proposed here has two parts. The first is to find the mix of agrotechnologies that best address the socioeconomic requirements. The second is to find the mix of agrotechnologies that take full ecological advantage of the physical, on-the-ground realities. The agrotechnologies, as the decision variables, are the same on both sides of the analysis. The departure from the norm lies not in finding a solution, but in finding ranges of feasible solutions.

The Objective Function

For the first part of the analysis, the socioeconomic objectives, i.e., profit, risk reduction, economic orientation, labor allocation, etc., are the linear programming constraints. For the second part, the actual land area, subdivided as to the physical farm characteristics, form the constraints.

For both parts or sides of the analysis, a single, nondescript or generic objective function is employed. This is

$$\text{Max}\ (T_1 + T_2 + \ldots + T_n)$$

where T_1 through T_n are the candidate agrotechnologies.

Solution Variables

For the different agrotechnologies, i.e., as the solution variables, some offer a clear starting point, others are mere sketches. Despite these shortcomings, like designs do share economic and ecological characteristics and it is fairly easy to assign element rankings. For the socioeconomic analysis, it is known, to a reasonable degree, which agrosystems are more profitable, which are cost effective, which handle risks better, and/or which allow for the better parceling of labor.

The same holds true for an area-classified farm. For this part of the analysis, there is also ample supporting data. The main concern, the erosion danger associated with the various agrotechnologies, has been quantified (Wiersum, 1984). The relative abundance of data extends into soil suitability for various crops.

Cross-plot or spillover ecology is not part of the initial assessment. Environmental issues, such as the ability to host regional fauna, might be included or deferred to later in the process.

An Applied Example

To demonstrate the optimization process, a hypothetical problem was formulated. Employing multiple objective linear programming (Steuer, 2009), this analysis looked at landscape with six candidate agrotechnologies as the solution variables.

Socio-Economic Solutions

Depending on the restrictiveness of the constraints, anywhere from five to fifteen efficient extreme points were found. From this range of solutions, four are shown in Table 12.1.

Table 12.1. Four possible land-use solutions for a given set of socioeconomic objectives.

Agrotechnology	Sample Solutions (% of farm area)			
Staple, first-tier risk monocrop	33.5	10.0	10.0	23.3
Staple, second-tier risk monocrop	10.0	10.0	50.0	10.0
Multi-species orchard	10.0	10.0	10.0	10.0
Shrub garden	0	47.5	0	0
Parkland pasture (third-tier risk)	15.0	15.0	30.0	0
Forest ecosystem	31.5	7.5	0	56.6

Land-Area Solutions

Figure 12.2 shows the overview of this hypothetical farm landscape. On the left is the unclassified or un-zoned landscape with elevational contours. The corners of the drawings represents the highest elevations. The idea is for fairly steep, but not excessive, topography. The farm buildings, upper center, orient the drawings.

In this hillside farm, the land is subdivided into three areas or zones. The middle is the most crop-favorable. The upper right corner represents a limited-use zone. The other areas, those shaded light gray, are steep and erosion prone. A more advanced version would fit crops and agrotechnologies to the soil characteristics, topography, and other site properties.

Again, employing the generic objective function, the three zones yielded about ten different land-use solutions.

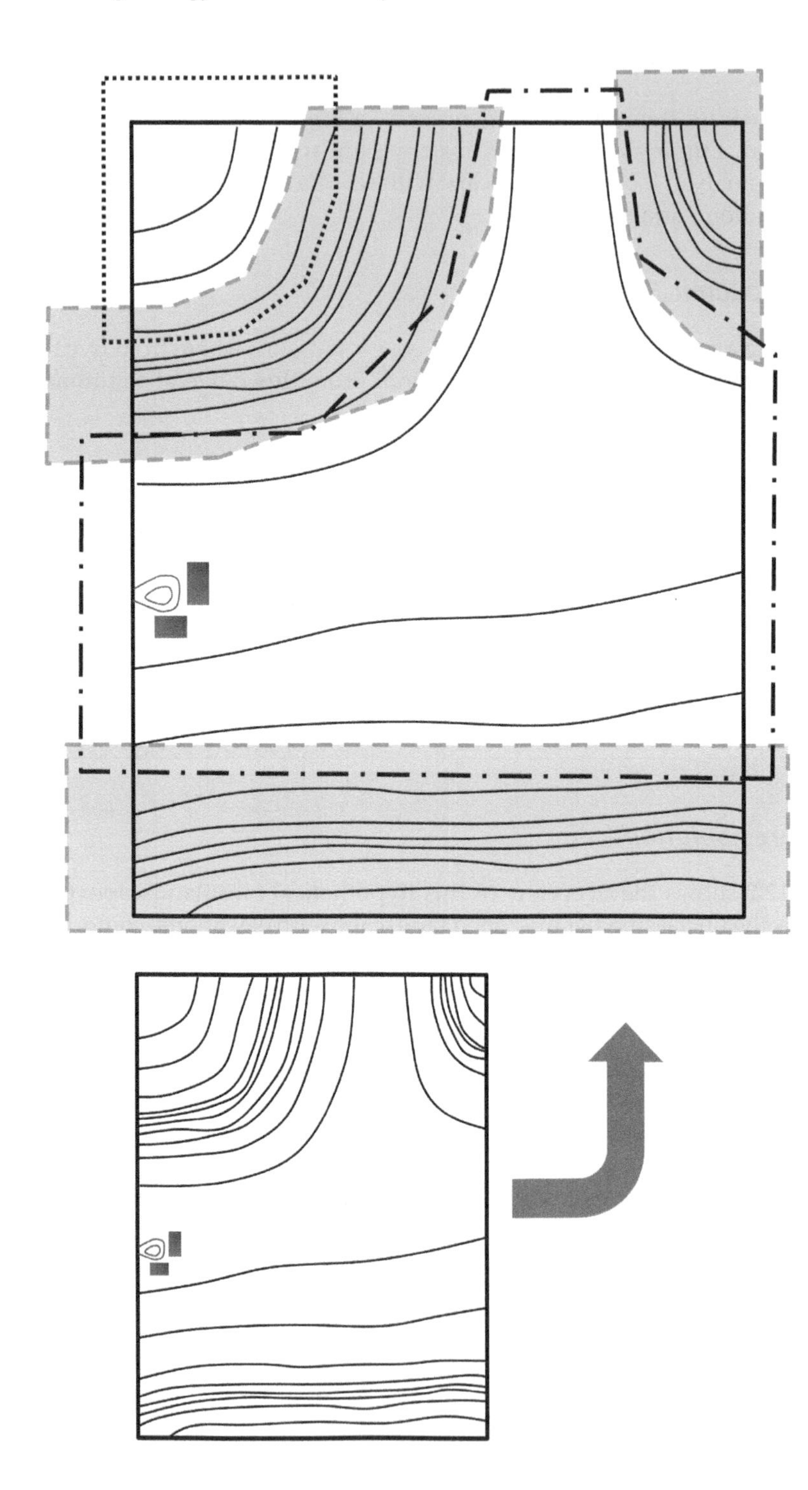

Figure 12.2. A farm landscape where the area is land-use classified.

Optimized

Optimization can occur where the two solution sets overlap or be some other interesting and relevant combination of the decision variables. After some manipulation of the constraints for both the socioeconomic (as in Table 12.1) and the zoned-land (Table 12.2) analyzes, a number of useful, but near-miss, solutions where found.

Some minor adjustments in the constraints produced one solution common to both sides of the analysis. Just to complete the process, the common solution is, in bold, the last column in Table 12.2. A loosening of the constraints would put more than one solution in the overlap of the two solution sets.

Table 12.2. Four possible land-use solutions for land-classified farm.

Agrotechnology	Sample Solutions (% of farm area)			
Staple, first-tier risk monocrop	30	40	40	**30**
Staple, second-tier risk monocrop	10	10	10	**10**
Multi-species orchard	0	0	0	**10**
Shrub garden	20	10	10	**10**
Parkland pasture (third-tier risk)	10	30	10	**10**
Forest ecosystem	30	10	30	**30**

A Final Layout

The one, arrived-at solution suggests a total area for each agrotechnology and an approximate location. The final layout is arrived at through a more detailed locational strategy, the rules of landscape design, and any practical considerations, e.g., the location of access roads.

Presumably, the crops, primary and secondary, for each candidate agrotechnology are known. This allows an assessment of the intra-plot ecology, the spillover and, therefore, the inter-plot ecology. The rules of the landscape design put this into practice.

Finding the by-area percentage for each solution variable/ agrotechnology and the relative location of each is the hardest step in the process. Beyond this, judgment and the rules for landscape design provide guidance for, and facilitate, the on-the-ground, final-fitting process.

The final step is to determine if the inclusion of bio-structures or land-modification agrotechnologies is warranted. In Figure 12.2, steep topography would indicate that infiltration ditches might prove advantageous.

Single Solution

One drawback is the reliance on uncommon software. For single objective linear programming, the approach would be to include the socioeconomic objectives as well as farm topography and soil properties as single-problem constraints. With the sample landscape (Figure 12.2), combining the two sets constraints into one problem gives, as expected, the same solution (Table 12.2, in bold).

With single-solution linear programming, a prodding of the objective will release additional outcomes. Despite these limitations, and the added work, having a respectable number of solutions from which to choose strengthens the process.

Multiple Solutions

There is a reason for a multiple-solution approach. In a procedure rife with limitations and non-quantifiables, this allows for an exploration, and selection, of those options beyond the single solution.

In Figure 12.2, the land-use zones overlap and each of these areas can be occupied by a range of agrotechnologies. This is part of the inexactness of landscape analysis.

As shown in Tables 12.1 and 12.2, the alternatives can be very diverse. Given that the constraints are approximations, the solutions represent tradeoffs and compromises. A loosening of the constraints produces more solutions and can increase the inherent flexibility.

It is also possible that none of the first-run solutions is suitable. Looking further afield might involve a slightly suboptimal solution or one found in one solution set (either in the socioeconomic or the field portion), but not the other.

Less optimal combinations can be favored when intangibles or quality-of-life needs are factored in. A scenic, aesthetically-pleasing landscape is outside the numbers and an example of a quality-of-life issue. A better balance of crops and forests could bring this about.

More is not always better. It is possible to become bogged down in solutions. This is less likely to happen if the number of solutions are adequate, but not overwhelming. Less that ten choices is considered manageable. Also, the whole process is made easier if one is not remotely involved, but has an intimate knowledge of the land and the socioeconomic requirements.

It would seem that the best starting situation would be a farm that is flat, fertile, and without obstructions. This is true to a small degree. Fewer solutions may be best on flat, soil-uniform sites where there are fewer intangibles. Where topography intercedes and/or the intangibles have a greater role, more solutions will insure a better fit. These patterns are part of solution theory.

13

Conclusion

Chapter Preview

Universal equations are a step forward, but are far from universal panacea. Agroecology is too complex for this to happen. There are additional topics which help order an expanded agroecology and the new paradigm that it represents.

Chapter Contents

Introduction

It has been said that, worldwide, very large farms consume 70% of the agricultural resources, i.e., seeds (purchased rather than farm raised), agrochemicals, etc., while producing 30% of the foodstuffs. For small farms, it is the reverse. They use 30% of the inputs while producing 70% of the food (Bitterman, 2013). On the surface, this may seem unbalanced, however, from an agroecological perspective, these are encouraging numbers.

Larger holdings, those of extensive monocultures and minimal agroecology, place a greater reliance on ex-farm inputs. Smaller entities would be more likely to control costs by extracting benefit from the inherent ecology. This means the more miserly and judicious use of ex-farm imports and better output/input ratios.

Photo 13.1. Artificially-maintained, industrial agriculture where crops are grown in an under-glass, controlled environment. This approach has a place, but it stands apart from those agrosystems that are intertwined with, and gain, from nature.

This helps illustrate the layers or complexity found in agroecology. This text has presented advanced concepts that explain the different expressions of biodiversity. With these in place, there are still questions on how agroecology is best manifested.

The Future

For some, moving from high-input agriculture to agroecology might represent only a small conceptional step. This is because there are levels of agroecology. These are independent of the reordering where agriculture, forestry, and agroforestry are positioned under the agroecological umbrella.

There is another aspect, that of the future (i.e., future directions) of agroecology. Given trends in monocropping, there exists impetus toward agroecology light. This would be minor moves away from chemical toxicity. The greater gains come with a willingness to embrace agroecology in all its dimensions, and with all its complexities.

Beyond Toxicity

In moving beyond the chemical dependency of the mainstay green revolution model, organic agriculture dispenses with the environmentally worrisome chemicals and forgoes the GM crops. Cropping versions of the factory farm can, under many existing regulations, be certified organic. As such, organic agriculture can be a variant of the green revolution model.

There are situations where large-expanse monocultures are hard to replace. Photo 13.2 shows a large area devoted to rice cultivation. Due to the presence of multiple land owners and a large, per-area population, this might not go beyond the single crop. This does not mean ecological improvements are not possible. Multi-varietalism remains a possibility.

With non-toxic inputs, ecological gains can accrue. Non-toxicity can encourage naturally-occurring bacteria and fungi and, hopefully, thriving and beneficial micro-communities. A lot of this relates specifically to thriving in-soil ecology.

This small step, organic agriculture and the non-use of toxic chemicals, opens the door for the widespread use of imported microbes and introduced predator insects. Although the latter may lack the habitat to fully establish, they can be artificially inserted for a short, rewarding stay. No longer poisoned and attracted by insect prey, birds and bats might also contribute.

The question asked in the first chapter, whether agroecology can feed the world, has been partially answered. This is certainly true with regard to organic agriculture (Badgley et al., 2007). The above options only add to this. This would be agroecology light.

Sustainable Agriculture

Further along this road, sustainable farming is roughly defined as the capacity of the land to yield into the distant future. Here the focus is on the temporal dimension.

Loss of productive potential can be from mismanagement which results in land degradation. This can also be caused by an inability to find critical inputs. Active and economically beneficial ecology will meet the generally-expressed criteria for sustainability. Although there are no set levels or standards, this can be, through the greater application of biodiversity, an improvement.

Photo 13.2. A large area rice monoculture. Because this a key staple crop and the land is subdivided by multiple owners, there is low likelihood for any major increase in biodiversity. The best outcome would be an organic, non-chemical future. The panorama is from Bangladesh.

Low/Mid-Level Agroecology

At the very least, the goal should be to bring large-scale monocultures more in agreement with, and more welcoming to, local flora and fauna. This might be rectified by positioning areas of natural vegetation or installing landscape structures, e.g., windbreaks. Through these, inter-plot ecological spillover can add a degree of natural dynamism to the yielding crops.

Better yet, it would be nice to expand upon cropping biodiversity. This is accomplished by mixing smaller, interacting plots, intercrops, rotations, or areas of natural vegetation. Full-capability, fully-fledged agroecology would employ all these options.

There is another aspect to mid-level agroecology. This would be the use of the general rules advanced in this book. These apply to, e.g., non-woody intercrops, complex woody agrosystems, and farm landscapes. The rules are user friendly, they sidestep a lot of agroecology details, and they deflect more complex analysis. Given the advantages, further development is warranted. This may be a necessary step and an objective (and a distillate) of the research process.

Full-Strength Agroecology

In full-fledged, full-strength agroecology, the reliance upon outside inputs is greatly reduced or entirely eliminated. The goal is, within the plot and the landscape, to extract as much as possible from what nature has on offer. The gains should translate into broadly-stated economic improvements. As emphasized, biodiversity, in all forms and manifestations, underlies all of agroecology.

As demonstrated throughout this text, biodiversity is not a linear function. The complexity associated with biodiversity grows exponentially with regard to number of species both of flora and of fauna.

The agroecological matrix is the key analytical tool. From this, it is possible to partition complex agroecosystems into their ecological parts. This should allow better use of the internal dynamics.

Another analytical tool is formulated for complex agroecosystems. Although this assumes that biodiversity is self governed (by way of density, diversity, disarray, and duration). This should allow inroads into the untapped potential of the agroforest.

With full-strength agroecology, outside inputs, are used only to rectify deficiencies and/or to provide short-term direction. If well done, yields will be good and costs low. The practical implications should be profitable, sustainable, risk-free, environmentally-friendly farming.

Agroecological Rankings

It is possible to measure or gauge the amount of agroecology present. At its best, agroecology is based upon the plentiful application of ecological dynamics. As the umbrella discipline, the amount of ecology ranges from the almost nonexistent to a total reliance. This is measurable through the agroecological matrix (as in Table 7.1, p. 100). Comparative rankings can find use, either for legal or for research purposes.

The agroecological matrix has two features. These are (1) the one-on-one and (2) the multipurpose counters. This allows for two subdivisions.

Agroecological strength involves an inter-matrix comparison of the actuated elements. In essence, the more non-zero elements, the stronger ecology. A simple addition of the elements in each, put into a ratio (i.e., the sum of the multipurpose elements divided by the sum of the one-on-one elements) would provide an apt measure of agroecological strength and/or intensity.

Agrotechnologies

This text has stressed the importance of the agrotechnologies. Most of the need resided in being a system of classification and eliminating the need to describe cropping systems in term of species being raised.

As expressions of biodiversity, agrotechnologies are often the vehicle, formally or informally, to capture and use set ecological dynamics for economic gain. These are many in number (Appendices 3 and 4), and in their use, their underlying ecology, and in their economics orientation. If this is not confusing enough, there are those management options/design vectors that can completely change the economic outlook, and even the use, of a given agrotechnology.

It helps that certain crops (i.e., seasonal, perennial, and/or woody) often have a climate-based associations with one or two agrotechnologies. This does not mean that, upon species selection, this line of development ends.

There can exist expressions of biodiversity that locals farmers have not yet seen, but could confer economic advantage. Also, there are a lot of unexplored and undocumented options, management or otherwise, that can better the ecological and economic results. Part of agroecology, as a science, is to provide this insight.

Analysis

For those less quantitatively inclined, rules/guidelines (e.g., for intercropping, complex agroecosystems, landscapes, etc.) are a complexity-reducing intermediate step. These allow productive agronomy to proceed without getting enmeshed in data, analysis, and ultimately quantified theory.

When dealing with any agroecological situation, (1) first look to the rules/guidelines and, if the question is not answered, (2) further analysis might be in order. This text introduces analytical techniques and equations that span a large percentage of the cropping situations encountered. Lacking much of the supporting data, the questions must be in a general form where generic functions provide a serviceable answer. For many, this is not exciting stuff.

Missing Dimensions

The focus of this text has been flora-centered agroecology. This is the most prevalent expression. Fauna-centered agroeocology is also a possibility where such systems exist. These can, and often, co-exist with flora-centered agrosystems.

Briefly mentioned are the ecological gains from birds, domestic and natural, in cropping environments. With this text, pastures are discussed mainly where forage, the primary crop, is directly harvested, i.e., grazed, by domestic animals.

There is an aspect to full-capability ecology where domestic birds and a range of domestic animals are both the profit center and the ecological nexus. These are farm-based systems. With this fauna-centric approach, bio-complex tree, grass, and shrub pastures are use managed and use maximized by rotations, not only with plants, but by sequencing domestic fowl and/or domestic animals, e.g., Strom (2013a).

Biodiverse fauna systems tend to mimic nature. For example, pigs raised in an agroforest and eating the fruits and nuts falling from trees is not that different from what occurs in a natural forest setting. The main difference is that the agroforest is more productive environment than many natural systems.

Other applications use the temporal dynamics of foraging flocks and grazing herds as they pass across the farm landscape. Throughout the world, there are large areas of grassland where grazing animals are the norm. For large ranches and free-range areas, the agroecology can be quite different and quite specialized.

There are management issues where the aim is not to degrade the productive capacity of the land. It is very possible, through herd management to increase the grazing capacity and the stocking rate. This is done through herd size, movement and even though the presence of predators (the idea is that predators, e.g., lions or wolfs, bunch herd animals to ecological benefit). This is part of grazing design where natural fauna is present. Without going into greater detail, a fauna-centric landscape is a specialized topic outside the scope of the crop-centered arguments in this text.

With both fauna and crop-centered landscapes, it helps if these are a mix agrosystems, including those thought of as forestry or agroforestry based. One reason lies in not overlooking any opportunities to expand upon the agro-possibilities and the agroecology.

As a Science

There are prerequisites to becoming a hard science replete with theory and mathematical arguments. With the exception of the empirically-based, land-use sciences, i.e., the non-agroecological versions of agriculture, forestry, and agroforestry, all sciences benefit from a strong theoretical, mathematically-derived base. As stated, this can be critical in that it provides the underpinnings from which to proceed. This is needed where experimentation is costly, complex, and/or extensive. All these conditions apply in an expanded agroecology.

In agroecology, empiricism is replaced with solid mathematical development. The previous chapters have outlined a framework that encompasses, if not all, most of agroecology. The notion being that, if theory is carried far enough, if quantified to a sufficient degree, and if not overly detailed, it conveys the essence of agroecology.

Theory is not an endless parade of ideas and concepts. Endpoints are reached. Over the decades, cropping models, in their attempts to mirror nature, present an implicit view as to how agrosystems function. Again, if concise and encompassing enough (and if theoretically, not empirically based), these can rise to the level of universal methodology and, by extension, a universal theorem. None have gone this far.

Agroecology differs in that the nucleus is not one or a few numerically-stated equations. Instead, it employs mathematical programming with one or more objective functions and a series of constraints. The constraints are situation dependent and set by the species, ecosystem, site, and other variables. These are added as needed.

For most plot/cropping applications, these constraints come by way of the agroecological matrix. For landscape design, the primary constraints are socioeconomic.

Having theoretical endpoints in the form of solvable equations untangles the base ecology and, in doing so, opens up, for exploitation, the underlying agro-dynamics. This is the very least that can be expected. With enough confidence, this becomes a (or the) core theorem. This is a strong statement and will be part of the debate as agroecology is further defined.

Sustainability Defined

A lessor agroecological definition is based upon sustainability. This is

> '*... an integral system of plant and animal productive practices having site-specific application but will last over the long term*' (Wikipedia, Sustainable Agriculture, 2013).

This implies that there is strong ecological underpinnings that promote durable ecosystems.

Contained is the notion that outputs are more or less certain despite local and/or regional upheavals. The above definition might be modified as

> *an integral system of plant and animal productive practices having site-specific application but will last over the long term and will continue provide for human needs despite changes in the political, social, and/or climatic situation.*

Part of insuring continued yields despite external strife lies in keeping the reasons for the yields close to home. This is done by utilizing farm or near-farm waste as nutrient sources. This is measured by the output/input ratio (Chapter 3). Part also lies in adopting climatic-based, risk-reduction measures (Chapter 10). Although not part of any definition, it is better that sustainable agriculture be inclusive in agroecology.

Agroecology Defined

In the first chapter, agroecology is presented as an umbrella discipline that, theoretically and conceptually, regularizes forestry and agroforestry. This reordering is structurally sound as trees impart a significant contribution to the ecology of both plots and farm landscapes.

By extension, agroecology includes organic, low-input, sustainable, permaculture and other variations whether the be within or away from the agronomic mainstream. As subdivisions, many will eventually be absorbed, without much residual trace, into the body of agroecological thought.

Given the current thinking, a definition of agroecology is a work in progress. There have been suggestions. For example, agroecology is

> '... *the study of the ecological processes that operate in agricultural production systems*' (Wikipedia, Agroecology, 2013).

Also possible as agroecology as

> '*the science of ecology applied to the design, development, and management of agriculture*' (Dictionary.com, Agroecology, 2013).

Since agroecology is replacing agriculture, the definitions should not differ. Agroecology (and agriculture) are therefore

> '*the cultivation of animals, plants, fungi, and other life forms for food, fiber, biofuels, drugs, and other products used to sustain and enhance human life*' (Wikipedia, Agriculture, 2013).

The view advanced in Chapter 1 has agriculture, forestry, and agroforestry as subsets of agroecology. When this view eventually prevails, this will sharpen the definitional debate. From this,

> *agroecology is the scientific basis for agriculture, forestry and agroforestry,*

or, with a complete fusion of the three subset sciences,

> *agroecology is the scientific (and ecological) basis for the cultivation of food, fuel, fiber, and other land-raised products with deference for and/or in cooperation with nature, natural processes, with the purpose of sustaining and enhancing of human life.*

Dichotomy

Cracks appear in trying to merge agriculture and agroecology. The green revolution model, specifically the factory farm, stands alone as threats are chemically thwarted, not defeated by countervailing ecology. Although the inputs can, and should, be organic, these systems are not ecologically supported, nor nature governed.

It can be said that two distinct agronomic sciences have emerged, (1) biodiversity-based agroecology and (2) monoculture-based agriculture. For some, a process underway. This is where the first approach seeks to reform the second.

Agriculture is the major land-use activity, taking 50% of the earths surface (news.nationalgeographic.com, 2005), and is the sustainer of human life. Most do not want large areas turned into an industrial landscape that is almost entirely devoid of natural processes and natural flora and fauna. The trend is in this direction where millions no longer connect, through farming, to the land. These are the social and political dimensions of agroecology. These arguments only touch upon the differences.

Precepts

There difference between agriculture and agroecology comes on many fronts. These separate agroecology from agriculture, both as practiced and in the directions taken. Summed here, these are

- where each crop is an element or component in a larger agroecosystem;
- where each plot is as part of larger, often natural, farm, local, or regional ecosystem that embraces inter-plot ecological solutions to plot-internal productive obstacles;
- through a longer time frame in planning and implementation;
- in a greater sensitivity toward and greater interest in diverse agricultural practices, one borne of an understanding of methods, applications, and the cultural role of agriculture;
- in the sanctity of the environment and with the do-no-harm admonition that accompanies the application of agroecological principles and practices;
- with an accentuation for improving the quality of life for the people and the natural fauna than inhabit the land;
- from an academic perspective, the elimination of those partitions that separate agriculture from forestry and agroforestry recognizing that trees are as much an ecological force in the farm landscape as are crops; and
- from the scientific perspective, in the wealth of theories, concepts, and principles from which to proceed when direct experience and research data is lacking (modified from Wojtkowski, 2006).

Non-Use

Being able to identify and measure the active (agro)ecology of plots and agroecosystems is one step. In many parts of the world, small-scale farmers have relied upon, and gained benefit from, various in-field and on-farm expressions of biodiversity. The hope is that this will continue.

In many developed countries, there seems some movement away from the chemically-dependent green revolution model onto fairly simple forms of organic agriculture. Full strength agroecology, one defined by flora and fauna biodiversity, should have, but has not, entered the picture. Among the reasons given (Wikipedia.com, Agroforestry, 2014) are the

- lack of developed markets for the alternate products
- unfamiliarity with agrotechnologies
- lack of awareness of successful examples

- not fully understanding how to use the competition and the gains from integrating trees, crops, and/or animals
- uncertainty as to the profit potential
- lack of demonstration sites
- expense of additional management
- lack of training or expertise, and/or technical assistance
- cannot afford adoption or start up costs, including the time needed
- unfamiliarity with marketing of the other products
- unavailability of field and farm level information
- apparent inconvenience
- lack of infrastructure (e.g., buildings, equipment)
- insufficient land
- lack of seed/seedling sources

Although the obstacles are many, these are not entirely insurmountable. Promoting full-on agroecology, replete with abundant biodiversity, may be the foremost mission. If active sites are scarce, farmers should have some exposure to the agro-alternatives. Further advancement can come through successful demonstrations.

To facilitate the process, any field approach must rely on established agrotechnologies. The rules/guidelines, as restated or developed in this text, allow for some site-based and/or user-needed modifications.

Agroecology

As the debate rages and trends become clear, the sciences advances. That put forth in the previous chapters focuses on the primary relationships and the primary inputs. This had led to what might be those apex algorithms that govern, through broadly started economic objectives, most of agroecology. Having an apex to the science should help focus the science.

Apex are only a base understanding. Agroecology is laden with nuances and exceptions. Whether these can be fully accounted for with sets of equations is an open question. Added to the burden are the cropping possibilities and options contained.

The proposed route is to flesh out the analytical algorithms, first with the possibilities and their options, later with the nuances and exceptions. This would be part of the long road to full-fledged, full-biodiversity agroecology.

A lot must transpire before this happens. It begins by accepting that biodiversity-rich agroecology can supply the worlds populations with food, fuel, fiber, and other farm and forest products.

Once this is accepted, there is little question that the 50% of the terrestrial lands, those that produce food for the many, need not be an environmental worry. Instead, well-formulated agroecological landscapes can safety secure those earth processes upon which human life, health, and happiness depends.

Appendix 1

Possible Equations, as Compiled from Various Sources

Allometric

$y = ax^b$ = a(x^b) where $0 < x < 1$

Bertatanffy

$Y = a(1 - e^{-bx})^3$ = a((1 – exp(–bx))^3)

Chapman-Richards

$Y = a(1 - e^{-bx})^c$ = a(1 – exp(–bx))^c

Gompertz

$Y = ae^{-be^{-cx}}$ = a exp(–b exp(–cx))

Hossfeld

$Y = \dfrac{x^c}{(b + \frac{x^c}{a})}$ = (x^c)/(b + ((x^c)/a))

Korf

$Y = ae^{-bx^{-c}}$ = a exp(–b(x^–c))

Levakovic (1)

$Y = a(\dfrac{x^d}{b + (x^d)})^c$ = a((x^d)/(b + (x^d)))^c

Levakovic (2)

$Y = a(\dfrac{x^2}{b + x^2})^c$ = a((x^2)/(b + (x^2)))^c

Logistic (1)

$$Y = \frac{a}{(1 + b\,e^{-cx})} = a/(1 + b \exp(-cx))$$

Logistic (2)

$$y = \frac{e^{a+bx}}{1 + e^{a+bx}} = \exp(a + bx)/(1 + (\exp(a + bx)))$$

Logistic (3)

$$Y = \frac{e^{a+bx}}{(1 + e^{a+bx})^2} = (\exp(a + bx))/((1 + (\exp(a + bx)))\hat{\ }2)$$

Negative exponential (Monomolecular)

$$Y = a(1 - b\,e^{-cx}) = a(1 - b \exp(-cx))$$

Sloboda

$$Y = ae^{-b\,e^{-cx^d}} = a(\exp(-b(\exp(-c(x\hat{\ }d)))))$$

Unnamed

$$Y = a(x^b)e^{(-cx)} = a(x\hat{\ }b)\exp(-cx)$$

Weibull

$$Y = a(1 - e^{-bx^c}) = a(1 - \exp(-b(x\hat{\ }c)))$$

Yoshida

$$Y = \left(\frac{a\,x^d}{b + x^d}\right) + c = ((a(x\hat{\ }d))/(b + (x\hat{\ }d))) + c$$

Appendix 2

Explanations Deferred

It would be a major diversion if every in-text technique and method mentioned required an in-chapter explanation. This is made difficult as much of the material has not been formally researched or the research is scanty. In addition, much of what is know on the methods and techniques is anecdotal, the result of infield observation, not based on rigorous science. Still, the techniques presented might prove of interest to readers. The hope is that some will take up the cause. An informal appendix might help in this regard.

The topics developed are

agrochemical alternatives
insect-eating birds
planting/pruning techniques

Safe, Effective, and Cheap Agrochemicals

Common soap—Some sources mention the need for a potassium fatty acid insecticidal soap (Wikipedia: Insecticidal Soap: 2013). It is not clear if this is important as many types have been used to good effect. The key to success seems in applying the soap solution as a rich, sudsy solution. This is sprayed directly on the targeted insects.

Those few studies that do exist show soap is clearly effective (Tremblay et al., 2004). Going beyond the research, this author finds that soap is quick killer. Death occurs within minutes, even seconds. The mechanisms are not fully enumerated, but include dehydration and asphyxiation. The latter may come by blocking the breathing tubes of the targeted insects.

This author has also found that soap works against some belowground, root-sucking insects, those that feed off the central root stem. For this, a small depression is dug around the plant stem above the targeted insects. This is filled with the soap solution. This may be best done when the soil is relatively dry and the soap solution can penetrate along the tap root.

Soap is generally used to counter insect outbreaks. The disadvantage is that it will kill predator insects as well as the destructive types.

Some farmers apply a soap solution every few days. It appears that soap has repellant properties where a coating on leaves deters insects. On small plots, the low cost allows frequent use. This author encountered this strategy in Africa.

Diatomaceous earth—This is applied as a sprayed powder. It functions by dehydrating insects. This is a contact killer with proven effectiveness (Wikipedia, Diatomaecous Earth, 2015). Diatomaceous earth is also a food additive. The only safety issue involves application, the sprayed or broadcast powder should not be inhaled.

This insecticide has a long history of non-use. For over a hundred years, generations of agronomists have mentioned the potential and suggested wide use. This has yet to happen.

Wood ash—Applied as a fine powder, ash keeps insects from feeding on plants. It may be best option as, in predator/prey setting, it does less harm to predator insects.

Neem oil—Extracted from the neem tree. Death comes after insects stop eating. It is sprayed on plants. It is not know if there is a secondary effect where predator insects die from eating herbivore insects that have consumed neem oil. This is safe, only a few minor health effects have been noted (Wikipedia: Neem Oil: 2013).

And against fungal diseases,

Milk—This is sprayed as a 5% milk, 95% water solution. This has been shown effective (Ferrando, 2007; Bettiol, 1999), but not against all types of fungi, e.g., Cromwell and Berkett (2009). Once more is known, this could be the equal of commercial insecticides, such as copper sulfate. This author found that spoiled milk was also efficacious.

Birds (employed to control insects)

(note: with no research, this is not published information. It was collected from those that have either used the techniques or regard it as viable.)

Domestic fowl—Farm fowl are the most time tested with a history of use that goes back hundreds, if not thousands, of years. Chickens may have been the primary species, but ducks and guinea hen have also found use in Africa and Asia. Within the last few hundred years, the turkey has been added to the mix.

Fowl in farm fields was so commonplace that the very early texts on agriculture only discuss those situations where birds are not crop suitable.

Lacking formal research, open questions are which species of birds are best in which situations. In addition, it would be nice to know the stocking rates and timing of their activity, e.g., when in the cropping sequence and at what times of day these are best employed.

The ecology is more complex than insect-seeking flocks of birds roaming farm fields. As these flocks approach, insects, to avoid being eaten, will take flight. In past eras, many of these insect would fall prey to farm-dwelling natural birds, those that capture flying insects. These include barn swallows and chimney swifts.

Barn swallows—These birds seek prey while in flight. Their habitat, open farm structures such as barns, makes these useful, but on smaller farms.

Chimney swifts—Paired with farm birds, chimney swifts have the most potential. These birds originally nested in standing hollow tree trunks, later finding habitat in unused chimneys. It is the unique nesting sites that make these birds potentially useful. Their numbers can be increased by establishing standing structures that resemble chimneys, e.g., large vertical tubes. This are placed, in number, around and in large plots. The advantage lies in having nesting structures that are not taken over by other, less-useful bird species.

Woodpeckers—This is a stand alone technique that does not require domestic fowl. Woodpeckers can be attracted to crop plots by establishing feeding trays. These are baited with peanut butter. If the populations of insects grow, the amount of peanut butter is lessened. The woodpeckers then seek nearby food. For best results, there should be nearby areas of standing, mature trees for nesting purpose.

Wrens—Beetles are a favorite food of wrens. The key is nearby wren houses. This requires management as other, unwanted and more aggressive bird species, the common sparrow, will take over the house. To be effective, the birdhouses must be actively emptied of other bird species.

Birds of prey—It can be useful to encourage rodent-eating and/or to discourage grain-eating birds. There is a traditional German technique that places widely-spaced, tall, T-shaped, roosting poles in farm fields. These are for hawks and owls. Part of this strategy would also include local nesting boxes.

Planting, Pruning, and Weeding Techniques

Planting—There are five planting methods. Most apply to trees. These are, with a brief explanation,

Seeds—plants are directly established from in-soil planted seeds.

Seedlings—Seeds are planted in a nursery or other site and, once established, they are moved to the final location.

Striplings—These are tree seedlings that are allowed to grow, in nursery, to height of 1.5 to 2 m, the side branches are pruned leaving a small, intact canopy. This is then transplanted to the final site. The purpose is to keep the tree from being destroyed by grazing animals after being moved.

Stems—This is where tree is grown to a great height, 5–10 m in a nursery setting. The main stem of the tree is cut above ground level and all branches are removed. The tall stem is then replanted. These can be very long, up to 10 m in length (as in Photo 9.2, p. 131). This method does not work with all tree species.

Stumps—For this, seedlings are raised in a nursery to height of 1 to 2 m. The trees are then pulled from the ground. The stem is cut about 5 cm above ground line, the root is cut about 10 to 15 cm below the ground line. The resulting short stump is pruned of side branches and side roots. This stump section is planted at the final site.

Photo A3. A stump section ready for planting.

Pruning—There are a number of pruning methods that are instrumental to the success of any agronomic endeavor. This list includes, again with a brief explanation,

Coppicing—This is were all stems of a plant are cut down to ground level. The notion is that this will regenerate into numerous stems. The purpose can be fencing or grazing.

Pollarding—This is where the stem of a tree is completely cut, usually at a height of 1.5 to 2 m. This can be to increase below-canopy light or to provide dry season forage to animals. The latter is cut and carried.

Stem pruning—The stems of trees are pruned of side branches, leaving a branch-free trunk with a reduced canopy. The purpose is to increase the value of the trunk as a salable commodity.

Lopping—With lopping, the outer fringes of a tree canopy are pruned, leaving a small, more compact canopy. The purpose may be to increase fruit set or to increase available below-canopy light.

Branch thinning—In order to increase the amount of light passing through the canopy, many of the secondary branched are removed. This leaves a more open canopy structure.

Weeding—As with all aspects of agroecology, there are associated options. For weed control, these include,

Total—For this, all unwanted plants are removed.

Select—Only those weeds that most effect the primary species are removed. Those weeds that do not influence growth and yield are allow to remain.

Near-plant—Only those that weeds that neighbor the primary species are removed. The others remain. This is common with trees where a fire close to the trunk can damage the tree. This can be part of an active fire strategy or a preventative measure.

In-row—Weeds near to the plant are allow to stay, those between the rows re-eliminated. This is done with weed-suppressing primary species knowing that, as the plant grows, the weeds close to the primary species will naturally succumb.

Appendix 3

The Agrotechnologies

- Productive agrotechnologies
  - Monocultural
    - Pure
    - Varietal/genus
  - Productive intercropping
    - Simple mixes
    - Strip cropping (seasonal)
    - Barrier or boundary
    - Complex agroecosystems (without trees)
  - Productive agroforestry
    - Isolated tree
    - Alley cropping (treerow)
    - Strip cropping (mixed tree)
    - Agroforestry intercropping
    - Shade systems (light)
    - Agroforests
  - Multi-species forest-tree plantations
  - Complex Agroecosystems
    - Natural pastures (highly biodiverse)
    - Mixed annuals
    - Intermediate polycultures
    - Agroforests
      - Tropical homegardens
      - Shrub gardens
      - Forest gardens
    - Enriched forests
      - Forest farming
      - Mixed cropping

        - Managed natural forests (silviculture)
            - Sparse and infrequent harvests
            - Damage-salvage cuttings
            - Species-oriented sequences
            - Senility cuttings
            - Selective-shelterwood thinnings
                - Light crown (the French method)
                - Heavy crown (the Danish method)
                - From below (the German method)
            - Liberation cuttings
                - Advancing
                - Reverting
- Facilitative agrotechnologies
    - Perceived monocultures
    - Facilitative intercropping
        - Simple mixes
        - Strip cropping
        - Boundary
        - Covercrops
    - Facilitative agroforestry
        - Parkland
        - Protective barrier
        - Alley cropping (hedgerow)
        - Strip cropping (woody)
        - Crop over tree
        - Physical support systems
        - Shade systems (heavy)
- Temporal agrotechnologies
    - Enduring or continual
    - Single rotations
    - Series rotations
    - Overlapping cycles
    - Taungyas
        - Simple
        - Extended
        - Multi-stage
        - End stage
    - Continual

- Feed systems
  - Pastoral
    - Pastures
    - Pastures with trees
    - Forage trees with pasture
    - Forage trees
  - Aqua-forestry
  - Entomo-agriculture
  - Avian-agriculture
  - Aqua-agriculture

- Other productive systems (floating gardens)

- Land modification
  - Erosion/infiltration
    - Absorption zones/micro-catchments
    - Gabons
    - Infiltration barriers
    - Mounds/beds
    - Buffer zones
    - Cajetes
    - Catchments
    - Riparian buffers
    - Stone clusters
    - Terraces
      - Tree based
      - Earthen
      - Stone
  - Water storage/transport
    - Paddies
    - Ponds
    - Water channels
    - Waterbreaks
- Bio-structures
  - Windbreaks/shelterbelts
  - Catchments
  - Anti-insect barriers
  - Corridors/habitats
  - Riparian buffers
  - Firebreaks
  - Living fences

Appendix 4

The Agrotechnologies Described

This section is an extension of agrotechnologies presented in Chapter 2. Previously, eight are discussed. These are

monocultures
intercropping
shade systems
alley cropping
agroforests
fallows taungyas
infiltration barriers
live fencing

Rather than provide complete descriptions of all those that have been identified (as listed in Appendix 3), these descriptions are deferred to this section. For those listed, the main headings are (1) productive, (2) facilitative, (3) temporal, (4) feed systems, (5) other productive, (6) land modification, and (7) bio-structures.

Productive Systems

Productive system as defined by competitive partitioning with little or no facilitative effects. For this category, all the component species offer economically interesting output. With these parameters, the systems can be subdivided into

monocultures
intercrops (seasonal)
tree/crop mixes (agroforestry)

multi-species forest-tree plantations
complex agroecosystems (agroforests)
feed systems

Monocultures

With one crop in one plot, this is an ecologically simple system. There are some variations, as

pure
varietal/genus (multi-varietal)

The monoculture is and will continue to be the mainstay of agriculture. For many crops, there are few, if any, alternatives. As a commercial species, sugarcane requires large areas and, although it can be intercropped, this is most likely not viable on a large-scale. Other commercial crops, wheat, sorghum, millet, if not intercropped, can be grow in alternating strips. This pattern confers ecological gains and reduces both biological and economic risk.

Another monocultural cropping alternative has more than one variety of the same crop mixed in the same plot. The multi-varietal monoculture offers many of the agroecological advantages of intercrop, e.g., reduced disease and protection against weather. This is a loosely defined grouping.

Normally this is a mix of diverse species that share the same scientific name. Thus, as varieties under the name *Brassica oleracea*, cabbage, Brussel sprouts, koribe, cauliflower, and broccoli would be intermixed. The alternative is to mix plants that share a common name but not a single scientific name. A site might have, different species of wheat, e.g., spelt (*Triticum aestivum*), einkorn (*T. monococcum*), durum (*T. durum*), khorasan (*T. turanicum*), etc. This also applies to forestry settings, e.g., where varieties of pine (*Pinus* spp.) share space with closely-related araucarias.

Intercrops (seasonal)

Many consider the Intercrop as a key expression of agroecology. There might be some truth in this as the cropping possibilities and accompanying gains are many. The categories are

simple mixes (seasonal intercropping)
strip cropping (seasonal)
barrier or boundary (productive)
complex agroecosystems (short duration)

Seasonal Intercrops

The intercrop is where, in the same space and time, one or more crop species are closely mixed. This number of possible crop combinations is large (as shown in the Table 6.1, p. 81). There are a number of agroecological advantages, e.g., better insect dynamics. Less disease spread, and greater per area yields. The main disadvantage is often the need for hand planting and hand harvesting.

Most of the simple mixes are revenue-oriented. There are a few cost-oriented intercrops. One example is wheat intercropped with cabbage. This is briefly presented in Chapter 6.

Strip Cropping

There are practical reasons for alternating crop strips. The strips can be, but not always are, monocultures. Even as single-crop strips, these offer some of the ecological advantages of an intercrop while mitigating some of the disadvantages. With strips, it is feasible to machine plant and harvest. Strips also alleviate some of the harvest problems associated with intercropping. One strip, usually monocultural, is harvested first. The now vacant strip provides room to efficiently harvest bordering intercrops.

The strips are usually 2–4 meters wide. These usually follow land contours, but can run straight if the topography is not severe. Generally, seasonal strips are revenue-oriented systems.

Boundary

There are boundary patterns that serve to protect crops. In most cases, these are composted of non-productive species. In a few examples, the boundary can be productive species. As these are generally taller than the non-boundary plants, the possibilities include sunflower, maize, and Jerusalem artichoke. Due to the awkwardness of a boundary harvest, these have less to offer in the way of wide-spread commercial application. A fully productive design is mostly confined to small-scale producers.

Complex Agroecosystems (short duration)

There exists a shorter term version of the complex agroecosystem. These contain a number of seasonal or short-duration species. The ecological gains can be quite large, but costs in planting and harvests mean that these have few commercial applications.

Tree/Crop Mixes

By definition, agroforestry is the interaction of multiple plant species including one or more tree species. These divide into productive and facilitative systems. For the productive versions, i.e., those composed totally of yielding species, the categories are

treerow alley cropping
strip cropping (tree/crop)
agroforestry intercropping
shade systems (with productive overstory)

Treerow Alley Cropping

Treerow alley cropping is a revenue-oriented agrosystem where a seasonal crop is raised between single rows of fruit or nut trees. These systems are not common as a flat site and a north-south row orientation are needed to insure that enough light reaches the crop. Also, not all fruit-bearing tree species are compatible with an accompanying cropping. The most common applications utilize, as the understory crop pasture grasses. These are either harvested or grazed.

Strip Cropping (tree/crop)

Strip cropping has more large-scale commercial potential. One strip will contain an annual crop, the alternating strip will be a fruit or nut-yielding tree species. There are some practical reasons. The trees can be harvested after the crop. This leaves space to pick the tree fruits. As with all these systems, the inter-row/strip ecological spillover can benefit both species.

Agroforestry Intercropping

Agroforestry intercropping is best described as a mixed-species orchard with many understory, often perennial species. These differ from the agroforest form in that these are ordered and spaced. A temperate version might mix apples, pears, peaches, plum, walnut, currents, and gooseberries.

The ordered and well spaced trees provide, in sum, the useful ecology and some economic gains. The fruits and nuts, harvested at different times, spread out the labor requirements, reducing the costs of large, temporary labor force. These are traditional in parts of Germany as streuobst systems.

Shade Systems (productive)

The most common of the shade systems are the facilitatory types. Less common are the productive versions. These differs in that both the under and overstory economically contribute. For these systems, the overstory spacing is wide enough so that the understory receives ample light. These differ from treerow alley cropping in that there is all the understory rows are shaded.

Multi-Species Forest-Tree Plantations

Not often found, but of great potential, are mixed species plantations. This would contain wood-producing species that normally grow in close proximity. These could be co-planted. This can be the equivalent of the natural forest except that it is ordered and spaced. The gains come in harvesting more wood per area. Where the inter-planted species mature at different times, it would be possible to get more wood in shorter time period.

Complex Agroecosystems

These are the topic of discussion in Chapter 11. Marked by high levels of biodiversity, there are distinct variations of intense biocomplexity. The groupings are

- natural pastures (highly biodiverse)
- mixed annuals
- intermediate polycultures
- agroforests
  - tropical homegardens
  - shrub gardens
  - forest gardens
- enriched forests
  - forest farming
  - mixed cropping
- managed natural forests (silviculture)

Natural Pastures

As complex agroecosystems, naturally growing pastures, with their mix of species, qualify as complex agrosystems. Because of their miniature scale, i.e., many small plants, these generally fall outside the management parameters associated with less dense, easier-to-manage complexity. Because of this, these are best categorized as feed systems.

Mixed Annuals

Complexity can be associated with garden plots. Growers can interplant many vegetable species and attain a high level of biodiversity. There are gains from the interplanting with less weeding, increased control of herbivore insects and diseases, and better utilization of essential resources. Since these agrosystems are mainly seasonal, the full gains from complex system may not mature, i.e., there is less time for the intra-plot ecology to reach a critical level. Being uncommon, these are a minor note and are more associated with backyard garden plots rather than having commercial applications.

Intermediate Polycultures

This is an agrotechnology first put forth in this text (Chapter 11). This is the logical halfway step between the intercrop and the agroforest. It contains more species than a plant-rich intercrop and fewer than the agroforest. The number ranges between about 5 to 12. These can be ordered or disarrayed. These systems are cost-oriented with both perennial tree and shrub components.

The envisioned use is as substitute for mono-species tree or treecrop plantations. The notion is to expand upon the biodiversity of these plantations for ecological gain and economic advantage. These would offer addition output with little sacrifice in primary output. The analytical methodology presented in Chapter 11 opens the door for commercial use.

Agroforests

Not to be confused with the broader agroforestry term, agroforests are species complex agroecosystem containing a large percentage of woody perennials. Agroforests represent a grouping of complex agroecosystems where woody perennials are intermixed with annual and/or perennial crops. The result is a system with high levels of biodiversity. A key elements are density, diversity, duration, and spatial disarray. This both simplifies management and maximizes the underlying, and beneficial, ecology.

In varying forms, agroforests are found throughout the humid tropical. Outside of this, these can be occasionally found in dryer regions. They do exist in temperate countries.

Homegardens

Homegardens are a version of the agroforest found close to or surrounding dwellings. These produce outputs destined for household consumption. This is the quintessential agroforest form. Located near populations, these have received the most attention.

Example are prevalent through the humid tropics. In some regions, these serve as adjunct agricultural systems, providing a wide range of household needs. The exception is often the staple crops. On other regions, these have greater importance, e.g., on many South Pacific islands, agroforests are the agricultural hub. This feature coconut, breadfruit, plantain/banana as well staple crops such as taro and yam.

There are other human beneficial dynamics associated with the homegarden. As an active ecosystem, these offer a good location to dispose of organic household waste. In decaying away, this provides input of mineral nutrients. This would be coupled with the use of the non-harvest option (where unpicked fruits are allowed to decay and recycle).

For some version, these contain ponds to raise fish for consumption. In yet another benefit, agroforests, located around houses, provide shade and cool localized climate.

Shrub Gardens

The shrub garden comes in two forms. One is a permanent landscape feature that, overtime, changes little as an intact agroecosystem. The second is transitional, evolving over a number of years into a taller agroforest, a forest garden, or into a forest/wood producing ecosystem.

For the permanent shrub garden, short-statured perennials, plus some annuals, would constitute the biodiversity. There are shorter trees, e.g., dragon fruit and the papaya, and dwarf varieties of temperate fruiting species. A temperate shrub garden might, for example, contain dwarf apple, pear, peach, and plum along with berries, e.g., currents, gooseberries, blueberries, and quince, and other perennials, e.g., horseradish, rhubarb, asparagus. Also intermixed would be some planted annuals, e.g., squash, climbing beans, and sunflower, are among the many possible biodiversity additions.

The transitional shrub garden is means from getting, productively, from one system to another. An annual monoculture or intercrop might be followed by some mix planting. As these grow, more species are added until, for a number of years, these form a shrub garden. A tropical example might be rice followed by cassava. As the cassava is harvested, shorter fruit or nut species are added. After these establish, the natural forest will be allowed to invade or the understory will be planted with one or more productive trees. The end result of this overlapping cycle could be a wood-producing natural forest, a forest garden, or a mixed species plantation.

Forest Gardens

More permanent and more specialized are the forest gardens. These are often designed around one to three primary species. Biodiversity does not require evenness in the number or the biomass of the individual species, there can be a bias for a few, usually but not always, taller trees. This is the basis for variation off the forest garden form.

The specialized systems takes advantage of this. This is a mix species plantation, wood, sap, fruit, and/or nut producing, that has these as the primary outputs. Within, and often below these trees, are diversity of other species. Most of the output and value is from the primary species. The other species, anywhere from 10 up to 100, provide the natural dynamics that ecologically maintain the forest garden.

Enriched Forests

In order to further reduce costs and get the system up and running quickly, farmers have the option of utilizing the trees on hand, i.e., enriching an existing forest. This can means directly replacing the understory of the natural forest with a perennial crop species. In tropical regions, coffee or cocoa are common. In temperate forests, ginseng or truffles, planted or encouraged under a forest canopy, are encountered.

Another type of an enriched forest is forest farming. The different lies in the output. With enriched forests, the emphasis is on the wood-yielding overstory. For the farmed forest, the economic strength lies in the understory crops. There are blocks or strips of managed forests or agroforests along with strip or blocks of crops. For example, tea can be in alternating strips with a shrub or taller agroforest. This would yield a mixed output with the most valuable and largest economic contribution coming from the tea component. This form, a clear primary species with multiple secondary outputs is possible with many perennial and annual crops.

Managed Natural Forests (Silviculture)

In order to accomplish both goals, wood and a welcoming natural ecosystem, silvicultural prescriptions have been developed. The forest itself can be considered the agrotechnology, the silvicultural prescriptions, i.e., the management method used, are the means. Briefly, these are

sparse and infrequent harvests
damage-salvage cuttings
species-oriented sequences
senility cuttings
selective-shelterwood thinnings
  light crown (the French method)
  heavy crown (the Danish method)
  from below (the German method)
liberation cuttings
  advancing
  reverting

With sparse and infrequent harvests, only a few trees are harvested and each of the harvests are decades apart. In order to do this profitably, only the most largest and most valuable trees are taken. None of the smaller trees are extracted. This can be found where there is only a plywood industry and large areas of remote forest. This is also found where the wood market, usually the export of logs, demands only large sizes.

With damage-salvage cuttings, the only wood harvested is that which has been damaged, i.e., wind toppled, killed by some disease, or fire harmed. If these events are infrequent, this management system will not provide enough wood to maintain a viable industry.

There are also species-oriented sequences. For this, each cutting or harvest cycle removes a different tree species. The cuttings can be at 10 year or greater intervals. With less species diverse forests, e.g., those in Europe, there would be limits on the size diameter. In highly species-diverse, tropical high forests, this impact would be far less.

Some feel that the best way to lessen the impact on forest ecosystem is to only cut only over-mature trees. Senility cuttings remove those that would die naturally in the next few years. A variation off this theme is to harvest mature trees, those that not longer experience high growth rates.

Among the commonly employed management systems are shelterwood thinnings. There are three variations, (1) light crown (also called the French method), (2) heavy crown (the Danish method), (3) from below (the German method).

The light canopy prescription first cuts larger, less value tree species with the goals of increasing the growth of the valuable species. Subsequent harvests still removes the smaller, less valuable trees, but also take a greater number of the now larger, valuable species.

The heavy crown system harvests only the largest size dominate trees. There is no limit on which species are cut. This opens the forest canopy for the faster understory growth, mostly in light demanding species. The goal is to maintain continued harvests of the largest trees.

The from-below method first cuts a large percentage of the lower suppressed tree, the next harvest is also of the suppressed tree species. After a few decades, what should be left are large diameter trees. As these larger diameter trees are cut, the system repeats, again starting with the suppressed understory.

In the long list of silvicultural prescriptions are liberation cuttings. This management method may be more suited to species-rich forest ecosystems. One requirement is that the forest have distinct succession phases, each with their own mix of species. This system has two variations, (1) advancing and (2) reverting.

When more of the valuable trees occur in the climax or ending stage of forest succession, it is economically beneficial to rush through the less valuable intermediate succession. The idea being is that, once the climax trees are in place as the understory, a harvest is undertaken. The purpose is to accelerate the growth of the more valuable climax species. Since the climax species are generally more shade tolerant, this normally a light overstory harvest.

If the reverse occurs, the most valuable trees are in an intermediate forest sucessional phase, it is advantageous to remove a fair percentage of the canopy. This should favor the growth of more light demanding, earlier successional species and result in a more valuable mix of forest tree species.

Other Productive Systems (Floating Gardens)

One highly specialized and a very intense form of agriculture is the raising of crops on floating mats or other buoyant structures. As minor agrotechnology, floating gardens are often constructed of bamboo or papyrus floats. These are covered with soil. Because these drift on rivers, ponds, and lakes, these provide a well-watered, but expensive, means to produce high value crops, economically are utilized where agricultural demands are high and where moist agricultural land is scarce.

Most of those mentioned are not floating gardens, e.g., as with the Aztec floating gardens, but describe farming along river banks. True examples are found in Bangladesh.

Facilitative Agrotechnologies

A large number of agrotechnologies are facilitative. These contain one of more secondary, non-productive plant species that have no economic purpose other than to help increase or maintain yields of the accompanying productive species.

As discussed in earlier chapters, what makes this a more involved topic is that there are different forms of facilitation. Not all involve essential resources. Generally, the best results come if the two component species are grown in close proximity, abutting or located in an adjoining strip or plot. For productive, LER-increasing facilitation, two categories are presented (1) crop and (2) agroforestry.

Crop Facilitation

There are inter-plantings where all the component species are annuals, but only one provides an economically useful output. The second species is facilitative.

- perceived 'monocultures'
- facilitative intercropping
  - simple mixes
  - strip cropping
  - boundary (facilitative)

A perceived monoculture has naturally occurring or planted covercrops growing with the crop. This may be thought of as a monoculture where the included species may offer significant, but unrecognized, facilitative ecology. The most notable are weeds. As an insect control, these normally unwanted plants can attract predator insects. Unacknowledged understory plants can be more common in orchard settings.

Simple mixes have often involve an understory covercrop. These can be multi-purpose. In addition to improving soil nitrogen (through nitrogen fixation), covercrops can repel harmful insects or provide protection through encouraged predator populations. Again, examples are many.

These same dynamics, minus the direct input of soil nutrients, are what give strip and boundary cropping their ecological advantages. Strips may be more ecologically interesting, and yield beneficial, if these are part of an active sequence of crop rotations.

Facilitative agroforestry

Pairing a productive with a non-productive species introduces cost issues. The facilitative species must provide enough inter-species value to cover the cost of planting and maintenance. This may not be justified if yearly replanting is needed. Costs are reduced if the non-productive facilitative species is a perennial. This is the reason for agroforestry-based facilitation. There are a number of possible formulations. These are

isolated tree
parkland
protective barrier
alley cropping (hedgerow)
strip cropping (woody)
crop over tree
physical support systems (vine over tree)
shade systems (with facilitative, usually dense shade)

Isolated Tree

A large, single tree located in a plot can constitute an agrotechnology if this tree has ecological and economic purpose. Also under this heading are designs that have two or three randomly placed trees. There is no restriction as to the type of tree, fruiting species, e.g., a large apple or mango tree, are often encountered. This contrasts with scattered trees of a parkland system where only specific, non-productive tree species are included.

Parkland

As expanded version of the isolated tree system, i.e., a single tree in large plot. Parkland systems contain multiple, widely-spaced, and non-arrayed trees of a single species. The defining characteristic, in addition to the dispersed tree placement, is the tree species present.

There are strict regional preferences with no more that two or three locally recognized species. The single species found in each parkland system would reflect the land purpose, i.e., crops, grazing, or crops in sequence with grazed fallow. The output from the trees, if any, is comparatively minor when compared against the value of the understory crop.

The natural equivalent is the grassy plains and the well-grazed savannas where scattered trees are present. The well grazed savannas of East Africa are the most recognized examples.

Comparatively well documented are the wide range of facilitative services offered. Along the in-soil gains are in a reduction in soil temperatures, greater nutrient content, faster water infiltration, low bulk density, and greater capacity to hold water.

Hedgerow Alley Cropping (facilitative)

Hedgerow alley cropping has potential in many regions where farmers operate on the margins. The idea is to replace long fallows with rows of trees between crop strips. Each year, the forage is cut and placed in the crop strip or placed and burned in the crop strip. This fertilizes a subsequent crop. The other alternative is a brief fallow where the trees grow and overtop the crop strip. Later the branches are cut, placed, and burned before planting begins.

This can be a significant nutrient gain. The highest recorded LER values, about 3.5, come from hedgerow alley cropping (Chapter 4).

Strip Cropping (woody)

Instead of having crops between single rows of trees, as with an alley system, the trees can be raised in wider strips. Within the tree strips, the rows are perpendicular to orientation of the strip. This allows the tree biomass to be easily cut and carried into the adjoining crops.

As an agrotechnology, strip cropping can alternate strips of annual crops or perennial species. The strip contents can be woody or non-woody. Also possible are facilitative strips and a nutrient beneficial inter-strip rotational pattern.

The primary reason for strip system lies along two fronts. The first is the contour placement and the anti-erosion, water-infiltration benefits of this layout. The second and lessor reason lies with insect control where this layout can discourse those insects that feed on specific crops.

Other than the above, gains come by way of special-purpose designs. These design are focus on the nutrient possibilities and, again, to a lesser extent on controlling unwanted insects.

Strip cropping may be the weaker cousin of alley cropping. Except for some gains in insect dynamics and erosion control, alley cropping seems the better choice.

Crop Over Tree

Most assume that the facilitative trees are taller than the crop. In a reversal of this, the crop can be taller than the tree. One variation off this has low pruned hedge over-topped by maize. On the larger scale, an orchard species can rise above a pruned facilitative tree. This is one of the less-utilized, less studies agrotechnologies with only a few published cases.

Physical Support Systems

Less utilized, vines can be grown over, and supported by trees. Tree-supported grapes have a long history, going back into very early, pre-Roman Europe. This involved grapes being held by short statured or highly pruned trees. This could find use and be reinstated if the ecology and economics are shown viable. Other vine over tree systems are found in recent times with pepper and kiwi fruit.

Facilitative Shade Systems

In contrast to the high-LER, high-revenue, light-shade systems, heavy-shade systems are common in the tropics. Coffee and cocoa being the primary crops. These systems are almost always cost oriented. These tend to be single output, that of the ground-level species. With the exception of an infrequent wood harvest, the overstory trees generally have little or no productive role.

The non-productive overstory provides a number of facilitative services. In cyclying nutrients, these should eliminate the need for fertilizers. Since productivity from the shaded understory is lower, less amounts of nutrients are withdrawn upon harvest, making these systems self-sustaining. The heavy canopy and in-place root structure keeps soils and essential nutrients from being eroded or washed away.

With two canopies, there is usually not enough light to support more than the two planned plant species. This means weeding costs are very low. With plenty of predator-prey relationships, insects losses should be minimal.

Temporal Sequences

An important aspect of agroecology is the temporal dimension (Chapter 9). These are not separate from spacial agrotechnologies as the temporal system link, in time, the spacial agrotechnologies. There are number of temporal strategies, these include

- fallows
    - natural
    - planted
    - productive

rotations
- single rotations
- series rotations
- overlapping cycles

taungyas
- simple
- extended
- multi-stage
- end stage

continual

Fallows

Fallows have existed since agriculture began. A period without crops and without the accompanying nutrient removal allows for an increase in the in-soil nutrients. A fallow period helps in that it permits, with high organic-content soils, an increase in the in-soil moisture content. To accomplish this, the crop-free interval can range from one season every few years to extended fallows lasting decades.

Other uses are in ridding the soil of unwanted insects, plant diseases, and weeds. This comes about through the change in the habitat ecology. In most cases, these organisms will be disappointed finding an unwelcoming ecosystems. This will break their reproductive cycle or they may fail to survive in the competitive, crop-free environment. This may not eliminate these as a future threat, but can reset their populations to near zero.

Some of the fallow options are discussed in Chapter 8. That presented here elaborates further on the options.

As expected, fallows come in various forms. This does not always relate to length, other factors define the type of fallow. These are (1) woody or non-woody, (2) burned or unburned, and (3) the species composition. Another category is the productive fallow.

An example is the classic fallow. For this, natural regrowth occurs, usually ending with small trees or woody shrubs. Before cropping, these are cut, allowed to dry, and then burned in place. Plowing is not always possible as stumps are a serious impediment but not always necessary as fire lays the soil bare for planting.

Other cases do much the same but, rather than burning at the end of the fallow, the leaves could be left in place to dry. After the leave shed, the branches are removed. There are a lot of obstacles to success with this variation, nutrient release timing, stumps that interfere with planting, and weed bloom that can follow post-fallow land clearing. The overriding advantage is the ability of the system to prevent soil loss through erosion. Hence, the main application is on erosion prone hillsides.

Another study is of a short fallow without woody regrowth. For this, grasses sprout anew from in-place root systems. The crop is planted before the grasses are burned. The fire, non intense, does not kill the crop. The grasses regrow during the cropping phase where the sprouting grasses undoubtedly help keep weeds at bay.

There are planted fallows were trees or a cover species is established during a cropping phase. This are cut, dried, and burned when the fallow ends. Later they reestablish or are replanted.

Productive fallows involves some harvest from a fallowed plot during the fallow period. The key requirement is that fewer nutrients are removed by the harvest than are replaced. With this limitation in mind, the most common example is a grazed pasture that exists only during a year long, inter-crop period. On-site animals redeposit nutrients, keeping the removal rate low. Despite the non-use of the land, fallow do have role even on intense agronomic settings.

Irrespective of the type of fallow, these come in various shapes. These can follow a block or strip pattern. There is also the temporal sequence. This is the order in which farm fields (whether strip or block patterned) are fallowed. This can go along with accompanying rotations of the crops.

Rotations

All agricultural and forestry species can be part of and benefit from a rotational sequence. These involve which crops to plant in a temporal sequence. The main reason for rotations is nutritional, using one planting to prepare the nutritional setting for an upcoming crop. This, plus any mineral that are added between seasons, should present the ideal balance of nutrients for an incoming crop.

There are a host of secondary reasons for rotations. These include ridding the land of certain in-soil herbivore insects and disease organisms. Rotations as also a lessor element in a strategy to reduce the populations and impact of weeds.

Within these parameters, there are single rotations were only a single seasonal overlap is considered. For example, a farmer might add marigolds one season as a nematode control for following crop. Also, bean might follow maize.

The second type is series rotation where one species follows another is a set order. This may continue, and include, many crop species.

Overlapping cycles are when one or more crops are planted before other crops are harvested. Usually this is accomplished with individual rows within one plot. Overlapping cycles are generally encountered in tropical regions. Continual systems those that have no planned ending but, throughout their life, there is a constant internal species renewal or replacement. This is found with agroforest-type systems.

Taungyas

Applied to tree-based systems, the idea of directing excess resources during one or more temporal phases to companion species is not new. This has been the norm across time and cultures. The term taungya originated in Burma during the early 1800s. It applied to teak plantings with farmer participation and a crop-aided establishment phase.

There are two ecological aspects of taungya use. The first is to utilize excess essential resources to raise a seasonal or short duration crop. This is done during one or more growth phase for orchards, treecrop plantations or in forest tree plantations. The second aspect of the taungya can be, but is not always, multi-participational, i.e., where other individuals or groups farm amongst the trees.

Taungyas exist because they confer economic advantage. In prime instances, revenue is increased as another crop is planted and harvested. As a secondary benefit, fertilization, applied to the crop, can accelerate tree growth. Costs are also reduced as weeds, those that would normally hinder tree growth, are controlled.

There are four taungya variations, (1) simple, (2) extended, (3) multi-stage, and (4) end stage. For the simple taungya, tree are planted they are generally small and drawing few essential resources. Weeds and/or erosion can be a problem. Weeding deals with one of the problems, covercrops and overplanting are answers to erosion. Both these solutions come with associated labor and other costs.

The extended taungya differs from a long-term, tree-based intercrop in that the understory continually changes with the different tree-growth phases. There can be anywhere from 2 to 6 cropping changes.

The first crop, planted when then the trees are small, are the most light demanding. There can be series of crops during this initial phase. As shade increases, this may be followed by longer term, more shade tolerant crop. The latter phase is often grazing with a shade resistant forage crop. After some years, the trees could be thinned and understory conditions might again favor light demanding short-duration crops. Latter stage thinning are found with some forest tree plantations.

The multi-stage taungya starts with plantation in some form, usually a multi-year fruiting species. As this reaches maturity, a second tree species is planted as the understory. Over time, the second planting replaces the first.

The initial crop can include species such as banana or papaya. Any fruiting or wood producing species can follow. For the second species, planting under an established crop can be disadvantage when shading retards growth or an advantage when being protected from drying in the hot sun.

The end-stage taungya are found with wood plantation where, as the trees mature, some are thinned. This many be done to encourage greater and faster growth in those few that remain. The now opened space, and surplus resources, can support crops or grazing.

Feed Systems

There exists a category of agrosystems where the focus is on fauna rather than crops. Although most of the variations are directed forage and grazing animals, there are systems for eatable insects, birds, fish, frogs, and larger mammals directly feed from the land. This is done through fauna-specialized systems.

Ponds can be established to raise fish, frogs, and ducks. Chickens, turkeys, and geese can roam fields or be fenced within feed-rich areas. Pastures of various forms are for cattle, goats, sheep, and other grazers. Pigs can be provided with species-diverse agroforest well stocked with fruit and nut-producing species or allowed time to browse within an oak/acorn parkland. Sericulture, the raising of silkworms, is an example of an insect-centered feed system.

There are cut-and-carry feed systems where the feed is harvested in one location and transported for use in another. Some harvest grasses to feed animals. The cut-and-carry systems are normally classified under some other system, e.g., forage crop alleys between rows of trees (treerow alley cropping).

For the direct feed systems, the basic categories are

- pastoral
  - pastures
  - pastures with trees
  - forage trees
  - forage trees with pasture
- aqua-agriculture
- aqua-forestry

entomo-agriculture
avian-agriculture

Pastures

Pastures are a common element in many agricultural settings. As repositories of beneficial insects, the ecology of a pasture contributes to the ecology of overall farm landscape. Although natural pastures fall under a complex agroecosystem heading, they are also classifiable under feed systems.

By far the most common direct-feed systems are those where cattle graze on ground-level vegetation. Common as this might be, there are many variations.

Cattle, as well as sheep, goats, pigs, geese, llamas, bison, and other animals, can be pastured. The forage can be directly eaten or the forage harvested for consumption at another time and/or place. As biodiverse, self-sustaining agroecosystems, most pastures are very cost oriented. Coupled with a low maintenance animal, these systems allow farmers to work larger areas than otherwise possible or to engage in very input-intense agriculture on other portions of the farm.

One saving is the harvest. This is reduced by allowing animals to directly graze. This is not always possible where grasses and other forage must be stored and provided during dry or through winter seasons.

There are a number of variations. The simple pastures is a common type of forage-based feed system. These contain multiple forage species.

Among possibilities is to utilize woody, short-statured, directly grazed perennials. Once established, these are low cost feed source whose primary advantage is drought resistance. These can convert dry pastures into a viable feed source even under severe conditions. An example is found in northern Chile where an Australian desert shrub species (*Altriplex nummularia*) replaces conventional grasses.

Rather than planting just shrubs or grasses, a mix of the two can be utilized. The common expression is to let the animals eat the ground-level forage. In the off season, the tree forage is consumed. This grass-shrub systems requires tree forage that is less desirable than the grasses and is eaten first. Later, then nothing else is available, animal partake of the shrub leaves.

The disadvantages of a shrub/grass design is the need for find a less palatable, but still nutritious shrub species. This may limit this system to one specific grazing animal.

A bit more versatile are tree-grass systems. Not quite as self-managing as shrub-tree systems, these require more labor input. For this, forage trees are planted above a pasture. These have taller trees and their leaves out of browsing harm. The grasses and other ground level forage are eaten first, after which the tree branches are hand-cut and supplied to the animals.

Aqua-Agriculture/Aqua-Forestry

This is a catch-all phrase for the raising of fish and/or other aquatic animals in a terrestrial farm landscape. This can include the raising of fish in large lakes, ponds, or rivers. Most common is the pond raising of fish, i.e., fish farming, as a commercial enterprise. Tilapia, catfish, carp are among the fish species that are raised.

In one form, fish can be added to rice paddies control mosquito larva, disrupt the life cycles of other waterborne insect pests, and as minor food source. Frogs can accomplish similar goals against flying insects. Aqua-agriculture can also be an addition in permanent irrigation ponds and in other active water channels.

It is possible to employ trees, directly or indirectly, as a feed source for aquatic creatures. Normally these would be fish, but frogs-supporting designs are also encountered. The advantage is that the fish, or frogs, are a cheap source of protein for farmers.

Entomo-Agriculture

The purposeful raising of insects in agricultural or forestry settings underlies entomo-agriculture. This can be more than insects as a supplementary addition. Some bugs can be economically important output. Honeybees with their service (pollination) and products (beeswax and honey) qualify. There are also many eatable insects that can be nutritious addition to the human diet.

Land Modification

There are a range of land changes that aid in the growing of crops. Most are directed to greater water capture and utilization through increased infiltration into the soil. The goals here is to slow or stop water from flowing along the surface and direct it into ground. Even when the soil is wet, land modifications are in place to prevent water-induced soil erosion. The range of modifications include

infiltration barriers
micro-catchments/absorption zones
terraces
 stone
 grass
 tree based
paddies
ponds
gabons
cajetes
stone clusters
stone fences
water channels
mounds and beds

Infiltration Barriers

To prevent erosion and to allow for surface water infiltration, it is possible to employ a ground cover and/or a barrier. The latter, under discussion here, are contour or cross-contour ditches, bunds, and/or hedges.

Infiltration barriers have a number of facilitative functions. Most, but not all, involve water management. Water, flowing below ground, confers a number of advantages over surface water. Among the gains, it does not contribute to erosion and it is available on site and to plants for a far longer period.

Ditches or bunds are mostly for water management. As hedges or through any accompanying vegetation that expands the ecological usefulness. Hedge barrier can double as fencing and/or as a refuges for insect-eating insects. Vegetation-filled ditches or vegetation-topped bunds maintain their effectiveness in water management, but also serve as to shelter carnivore insects.

Micro-Catchments/Absorption Zones

Rather than long infiltration ditches and/or hedges, it is possible to place, on hillsides, smaller diversion ditches. This are only a meter or two wide and generally divert rainfall to a small hole. Adjacent the small hole is the recipient plant, usually a tree or shrub.

The intended target of this effort would benefit from greater survival. Micro-catchments are generally employed in semi-arid regions or where rainfall is erratic. Although this is a more expensive way to grow trees, with costs up by about 25%, survival rates are expected to increase by 50% to 90%.

Micro-catchments can have landscape intent. These are used, closely spaced, for multiple tree plantings. The idea being that hillside runoff is eliminated along with any erosion threat. Also, the rainwater, now flowing under ground, is on site and available for a longer period.

Terraces

For agriculture on steep hillsides, the terrace is a common and necessary structure. Because of the associated productive and economic gains, these are familiar in many rural landscape. The importance of terraces on hillsides is without question. There role in erosion control is a principle reason for use. In some regions, those where the soil are less erosion prone, it is possible to farm hillsides without investing in terraces. This situation is found in the mountain regions of New Guinea. This is the exception.

In addition to preventing erosion, terraces increase the efficiency of farm work. It is easier to operate on flatland rather that trying to labor while clinging to a steep hillside.

There are a few other advantages. Cold air flows downward and terraces actively shed frosts. Stone construction helps as the heat absorbed during the day is released at night. This can keep a more favorable temperate during a cool night.

There are some other gains. Flat terraces are good infiltration structures and help keep the soil moist between periods of rainfall.

There are three design aspects to any terrace. These are the outward projection (how a terrace projects outward from the hillside), the side slope (level or an upward ramp type), and the facing (supported by stone, grass, or shrub or tree). The common, rice-paddy terrace would be flat, level, and grass faced.

Paddies

As one of the land-modification agrotechnologies, paddies are generally associated with the growing of rice. Cranberries and wild rice are also associated with swamp-like environment.

Paddies are important from two perspectives. The first is the popularity of rice as a staple crop. The second reason is that, due to the actual area devoted to the raising of rice, so much land is also put into supporting agrotechnologies, e.g., water catchments, irrigation channels, etc.

Ponds

As part of rural landscapes, ponds store water for irrigation, quench the thirst of farm animals, are the basis for aqua-agriculture, serve farm families and, among the other benefits, can help support natural fauna. The mainstay is the long-standing farm pond. In some countries, fire insurance, not agricultural usefulness, may be the principle reason for many ponds. Insurance is cheaper if fire-suppressing water is close at hand.

In some tropical regions, fish are a food source and ponds may help in this. Domestic fowl, such as duck and geese, benefit for these structures. In small farm landscape, nutrient-rich sludge, dug from the bottom of ponds, many be added to gardens.

Ponds that are part of a flood cycle can be natural source of fish, but less useful for commercial purpose. These temporary water bodies are of use as they keep water on site and help raise the water table.

Gabons

One of the many flood and drought defenses are rock filled wire baskets placed in wades and streams. These slow the water flow, promote infiltration, and keep soil from washing downstream. As with most water management structures, gabons serve a double purpose, controlling flows during periods of high rainfall and by keeping water from being lost, by encouraging in-soil absorption, during periods of drought.

The wire baskets, filled with local rocks, are placed across dry beds and active streams. These should not wash away when rainfall and water flows are high. The key aspect is to first place the gabons at higher elevations in watersheds. These are later established at the lower altitudes. This placement order insures that these are most effective in their intended use and will not be overwhelmed by surges of water. Ideally, these should spaced so that the water, when stopped or slowed up by one gabon, just touches the base of the next up-hill placement.

Cajetes

Rather than positioning long contour ditches as a means to capture, hold, and infiltrate water, this task can be done equally well through well positioned, individual ditches. Generally, these cajetes are spaced across hillsides to maximize water capture and are fairly deep in order to stop and contain high rainfall runoff.

As a land-modification agrotechnology, these designs are useful when contour ditches cannot be fully employed. The common situation is where boulders litter a hillside, making it impossible to establish long, continuous, contour structures.

These can be small, one-half meter across or be quite large, several meters across. Larger cajetes are placed along side wades or intermediate streams to capture runoff and allow time for this in infiltrate into the ground. These are not ponds in that they are expected to dry completely either during a dry season or between rains. Until drying does take place, these may function as water holes for animals.

Stone Clusters

Where stones are prevalent and the climate dry with short heavy rains, stones can support agriculture. This can be crops raised between unpiled rocks or small enclosures a few meters, e.g., two to five, in area. In this form, the rocks moderate daytime and nighttime temperatures, encourages rains to permeate the soil, and keep the soil from washing or blowing away. Over time, the physical, and chemical breakdown of rocks can supply mineral nutrients. These small plots enclosed by stone fencing are an unusual and only found in unique settings.

Stone Fences

Stone fencing can be agroecologically important or an imposing, but ecologically minor, landscape feature. As a minor feature, stone fencing is often a place to put the stones that would otherwise impede cultivation. In an expanded role, stone fencing can help with water infiltration. If the plot are small and bordered by stones, they can moderate temperature extremes. These are expensive to construct, but have very low yearly maintenance costs. The benefits of well placed stone fencing can accrue for years.

Water Channels

Among the land modification agrotechnologies are small canals or channels used to convey water across farm landscape. These can be a supporting feature carrying irrigation or household water or, within their own right, these can be an ecological and an economic feature within the farm landscape.

Channels are not wide, usually not more than one meter. These can be dry some of the year or wet year round.

Mounds and Beds

Raised soil structures, either as individual mounds or larger beds, can be diminutive, a few centimeters in height, or be as much as one meter tall. Mounds and beds provide two ecological services. These protect against low temperature extremes and keep plant roots out of waterlogged soils.

The temperature differences between top of a mound and the between-mound trough with only a few degrees, normally 2°. This is for a ½ m tall mound. Although seemingly minor difference, this can be enough to insure plant survival. These are found more in high mountain regions where nighttime temperatures can plunge low enough to damage crops, e.g., at altitude in the Andes and in New Guinea. This are also utilized where springtime temperatures, those that occur after planting, can plunge during night hours.

Also to counter frosts, mound can be constructed from soil covered biomass. The decaying vegetation with in the mound generates heat, further protecting a sensitive crop.

In addition, mounds find use where soils are or can become waterlogged. Some crops, maize being a prime example, will die if the roots are under water. Mounds prevent this.

Mounds can be placed in flooded fields. This design utilizes the heat-holding ability of water to combat low temperatures. The area between the mound, with free standing water, holds the daytime heat. The mound keeps the roots of crops from totally saturated.

Bio-Structures

Between productive plots and within an agronomic landscape are non-productive agrotechnologies. These can be, as above, the land modification agrotechnologies or a series of non-productive bio-structures. Together, these group as auxiliary agrotechnologies.

Whereas the land modification agrotechnologies manage water dynamics, bio-structures have a wider role and help with water, winds, harmful insects, plant diseases, small and large animals, and other cropping threats. The landscape bio-structures include

windbreaks/shelterbelts
waterbreaks
catchments
anti-insect barriers
corridors/habitats

riparian buffers
firebreaks
living fences

Windbreaks

A standard feature of many agricultural landscapes is the use of wind blocking tree barriers. These can be as windbreaks or, within large farms or across regions, a mix of windbreaks and shelterbelts. Windbreaks are only one to three trees wide shelterbelts are generally wider, more substantial in design, less closely spaced within a farm landscape.

In addition to their main role of blocking prevailing winds, windbreaks and shelterbelts accomplish a number of ecological tasks. A reduction in the wind does retard plant and soil drying. This can have a substantial effect upon yields. Windbreaks eliminate sandblasting, where windbourne sand particles strike and injure plant leaves and stems. The healing needed takes resources away from seed or fruit production.

The positive effects of a wind sheltered environment transfers to domestic animals. Being protected from hot or cold winds allows for faster and greater weight gain. Protection also reduces the mortality for the more susceptible, younger animals.

As a side benefit, windbreaks can serve as live fencing. In the off season, when crops are not in place, a reduction in winds prevents soil erosion.

Waterbreaks

As part of an anti-flood defense, strips of trees or shrubs can be placed across lands that flood. These are placed perpendicular to the water flow.

Bio-structures are employed when flooding is an irregular occurrence but, when it does happen, valuable soils are washed away, Waterbreaks prevent soil loss. If well designed and placed and under the right conditions, these may even encourage the deposit of nutrient rich silt. These are simple structures which can double as windbreaks and/or animal fencing.

Catchments

There can be areas, farm wide or regional, there the collection of water is more important than any productive role the land might have. These catchment zones are found where water must be diverted to another use. Examples are areas where water is captured for use in water-intensive rice paddies.

These catchment area can serve a productive function. Natural forests and forest tree or tree crop plantations, if well managed, will not effect water quality and water runoff. There are caveats, the tree species selected must not be thirsty. This would eliminate heavy water-using tree. Examples are thirsty, but drought resistant, species of eucalyptus. The liberal use of riparian buffers, infiltration ditches and other water management features will further purpose a catchment.

Anti-Insect Barriers

Strips of vegetation can be specifically placed to slow or halt the spread of damaging insects. These are usually biodiverse with flowering plants. The latter is to attract predator-type insects. These barriers are usually about one meter wide but where these have other purpose, e.g., as windbreak, other widths are possible.

Corridors/Habitats

Across landscapes, corridors permit freer travel by a range of beneficial organisms. The idea is that these will more from reservoirs and habitats to areas where their services are of need. Corridors also help with pollinating insects and larger animal that eat insects, e.g., skunks. These are also repositories of micro-flora and micro-fauna.

As less of an agricultural aid and more of an environmental service, wide corridors are also needed by wildlife species. These allow timid animals to travel, avoiding unwanted contract with people and predators, and for tree dwellers to reach their preferred habitat without crossing open ground. Windbreaks would offer this benefit.

Riparian Buffers

The primary purpose of the riparian buffer is to keep soil and nutrients from washing into and contaminating active watercourses. The most common form is a band of vegetation along the sides of active streams, rivers, lakes, or ponds. These can be two to four meters wide and can be wider where water flows are highest. This means that the protective vegetative strip can extend upward and outward, often to a great distance. This is often along a dry wade. Usually these buffer strips are composed of perennial, nutrient-demanding plants, those that are better at absorbing the in-water nutrients.

Firebreaks

In some regions, fire is a danger to agriculture and forestry. This danger applied to some annuals crops and tree plantations. The purpose of a fire barrier is to contain the spread of the danger.

The best barriers are of bare earth. These are expensive to maintain and are prone to erosion. Other types more agronomic. One is a root crop. These are in place when the fire danger is low, harvested before the fire season peaks. Once harvested, the strip would lack vegetation during the dryer, low-erosion period when the fire threat is greatest. Another options is a crop that does not easily burn. Cacti are often employed.

Living Fences

Fencing can be more than the partitioning of farms and farm landscapes. This can be, if the right type of fencing is utilized, an agroecologically-positive addition. Wire fencing, the current standard, is single purpose with no ecological role. Vegetative type have multiple uses within landscape. Besides fencing, these serve as infiltration structures, riparian buffers, for erosion control, as corridors for the movement of insect-eating insects, and have lessor purpose as windbreak.

There are the variations of the living fence where barbed wire is strung on live posts. One design has normal post spacing, a second is the same inter-post plan but with spiny vine growing on the wire. The vine-covered wire design can be utilized with dead posts but is better with live posts. This is because the vines will impede post replacement.

In yet another wire-with-live-posts design, there is the possibility for a wider inter-post spacing. For this, woody branches are interwoven into the wire. These branches can be the branch sprouts that grow and are cut from the posts.

The other live fence types are all vegetative. There is the open hedge where animal unfriendly plants are closely spaced. The classic example is the aforementioned cacti example. Other plants, including sisal and, in Europe, the white thorn, are used in these open hedge designs.

The pleached hedge uses fast establishing tree or shrub with interwoven branches. This forms a dense, impenetrable barrier. Usually the branches sticking out from the barrier are cut away, producing a two dimensional structure. The cut branches are placed along the fence to close any below-barrier openings. Further strengthening the underside, some suggest planting the hedge on the top of a small bund. This suggestion may also involve survival and uniform growth.

For the plashed hedge, saplings are planted or allowed to grow. These are spaced every few, i.e., 2–3, meter. When they reach a ground-level diameter of about 4–5 centimeter, these are cut, at ground level, midway through, bent and staked to the earth. The result would be an upward branch growth and downward rooting. This produces a uniform hedge that, for further strength, can be pleached.

References

Adeyemi, O., Fabuumi, T.O., Adedeji, V.O. and Adigun, A.J. (2014) Effects of time of weed removal and cropping system on weed control and crop performance in okra/ Amaranthus intercrop. *American Journal of Experimental Agriculture* 4(12): 1697–1707.

Altieri, M.A. (1995) *Agroecology: The Science of Sustainable Agriculture,* Westview Press, Colorado, 433p.

Altieri, M.A. and Nicholls, C.I. (2004) *Biodiversity and pest management in agroecosystems,* Food Products Press, NY.

Anderson, R.L. (2003) An ecological approach to strengthen weed management in the semiarid Great Plains. *Advances in Agronomy* 80: 33–62.

Anderson, R.L. (2005) A multi-tactic approach to manage weed population dynamics in crop rotations. *Agronomy Journal* 97(6): 1579–1583.

Badgley, C., Moghtader, J., Quintero, E., Zaken, E., Chappell, M.J., Avilés-Vázquez, K., Samulon, A. and Perfecto, I. (2007) Organic agriculture and the global food supply. *Renewable Agriculture and Food Systems* 22(2): 86–108.

Bardner, R. and Fletcher, K.E. (1971) Insect infestations and their effect on the growth and yield of field crops: a review. *Bulletin of Entomological Research* 64(1): 141–160.

Baskin, Y. (1994) Ecologists dare to ask: how much does diversity matter? *Science* 264(8 April – 5156): 202–203.

Berek, P. and Helfand, G. (1990) Reconciling the von Liebig and differential crop production functions. *American Journal of Agricultural Economics* 72(4): 985–996.

Bettiol, W. (1999) Effectiveness of cow's milk against zucchini squash powdery mildew (*Sphaerotheca fuliginea*) in greenhouse conditions. *Crop Protection* 18(8): 489–492.

Bitterman, M (2013) How to feed the world. *The New York Times, Sunday Review* CLXIII(56,295-Oct. 20): 9.

Bolfrey-Arku, G.E.-K., Onokpise, O.U., Carson, A.G. Shilling, D.G. and Coultas, C.C. (2006) The speargrass (*Imperata cylindrica* (L) Beauv.) menace in Ghana: Incidence, farmer perceptions and control practices in the forest and forest-savanna transition agro-ecological zones of Ghana. *West Africa Journal of Applied Ecology* 10(1): African Journals On Line.

Birrer, S., Zellweger-Fischer, J., Stoeckli, S., Kormer-Mevergelt, F., Balmen, O., Jenny, M. and Pfiffnor, L. (2014) Biodiversity at the farm scale: A novel credit point system. *Agriculture, Ecosystems and Environment* 197: 195–203.

Brown, M.W. and Tworkoski, T. (2004) Pest management benefits of compost mulch in apple orchards. *Agriculture, Ecosystems and Environment* 103(3): 465–472.

Caborn, J.M. (1965) *Shelterbelts and Windbreaks,* Faber and Faber, London, UK, 287p.

Coffey, K. (2002) Quantitative methods for the analysis of agrobiodiversity. p. 78–95. In: Brookfield, H., Padoch, C., Parsons, H. and Stocking, M. (eds.) *Cultivating Biodiversity:*

Understanding, Analyzing, and Using Agricultural Biodiversity, ITDG Publishing, London, 292p.

Cousens, R. (1985) A simple model relating yield loss to weed density. *Annals of Applied Botany* 107: 239–252.

Cromwell, M. and Berkett, L. (2009) *Summary of a Preliminary Evaluation of Raw Milk as a Fungicide for Apple Scab Management in Vermont,* on-line: uvm.com.

De Moura, E.G., Sobrinho, J.R., Aguilar, A.D., Serpa, S.S. and dos Santos, J.G. (2010) Nutrient use efficiency is alley cropping systems in the Amazonian periphery. *Plant and Soil* 335(1-2): 363–371.

De Schutler, O. (2012) Agroecology, a tool for the realization of the right to food. p. 1–23. In: Lichfouse, E. (ed.) *Agroecology and Strategies for Climate Change,* Springer, NY, 333p.

De Vrieze, J. (2015) The littlest farmhand. *Science* 349(14 August-6649): 680–683.

Ferrando, F.J. (2007) The effect of milk-based foliar sprays on yield components of field pumpkins with powdery mildew. *Crop Protection* 26(4): 657–663.

Gaiud, S. (2011) Microbial Inoculant: an Approach to Sustainable Agriculture, on-line: BiotechArticles.com.

Glover, J. (1957) The relationship between total seasonal rainfall and yield of maize in the Kenya highlands. *Journal of Agricultural Science* 49: 285–290.

Griffin, R.C., Montgomery, J.M. and Rister, M.F. (1987) Selecting functional forms in production function analysis. *Western Journal of Agricultural Economics* 12(2): 216–227.

Grimm, S.S., Paris, Q. and Williams, W.A. (1987) A von Liebig model for water and nitrogen crop response. *Western Journal of Agricultural Economics* 12(2): 182–192.

Harrison, G. (1775) *Agriculture Delineated,* J. Wilkie, London, 414p.

Harmsen, K. (2000) A modified Mitscherlich equation for rainfall crop production in semi-arid areas: 1 Theory. *Netherlands Journal of Agricultural Science* 48: 237–250.

Hiebsch, C.K. and McCollum, R.E. (1987) Area-x-time equivalency ratio: A method for evaluating the productivity of intercrops. *Agronomy Journal* 70: 15–22.

Hillel, D. (2004) *Introduction to Environmental Soil Physics,* Elsevier, Amsterdam, 494p.

Holford, I.C.R., Doyle, A.D. and Leckie, C.C. (1992) Nitrogen response characteristics of wheat protein in relation to yield response and the interaction with phosphorus. *Australian Journal of Agriculture Research* 43(5): 969–986.

Jensen, M. (1993) Soil conditions, vegetation structure and biomass of a Javanese homegarden. *Agroforestry Systems* 24: 171–186.

Kamani, J.N. (1987) *An Agronomic Evaluation of Potato/Maize and Potato/Bean Intercrop,* on-line: University of Nairobi Digital Repository.

Kareiva, P. (1994) Diversity begets productivity. *Nature* 368 (21 April): 686–687.

King. K.F.S. (1968) Agri-Silviculture, The Taungya System, Bulletin No. 1, Dept. of Forestry, University of Ibadan, Nigeria, 109p.

Llewelyn, R.V. and Featherstone, A.M. (1997) A comparison of crop production using simulated data for irrigated corn in western Kansas. *Agricultural Systems* 54(4): 521–538.

Loudon, J.C. (1826) *Encyclopedia of Agriculture,* Longman, Hurst, Rees, Orme, Brown, and Green, London, 1226p.

Malcolm, J.R. (1994) Edge effects in central Amazonian forest fragments. *Ecology* 75(8): 2438–2445.

Mathews, B.W. and Hopkins, K.D. (1999) Superiority of S-shaped (sigmoidal) yield response curves for explaining low level nitrogen and phosphorus fertilization responses in the humid tropics. *Journal of Hawaiian Pacific Agriculture* 10: 33–46.

Mead, R. and Willey, J. (1980) The concept of the 'land equivalent ratio' and the yield advantages from intercropping. *Experimental Agriculture* 16: 217–228.

Morales, H., Perfecto, I. and Ferguson, B. (2001) Traditional fertilization and its effect on corn insect populations in the Guatemalan highlands. *Agriculture, Ecosystems and Environment* 84: 145–155.

Morris, R.A. and Garrity, D.P. (1993a) Resource capture and utilization in intercropping: non-nitrogen nutrients. *Field Crops Research* 34: 319–334.

Morris, R.A. and Garrity, D.P. (1993b) Resource capture and utilization in intercropping: water. *Field Crops Research* 34: 303–317.

Muuisse, P., Jensen, B.D., Quilambo, O.A., Andersen, S.B. and Christiansen, J.I. (2012) Watermelon intercropped with cereals in semi-arid conditions: An on-farm study. *Experimental Agriculture* 48(3): 388–398.

Nasser, R., Velásquez, C., Velasco, C., Ruíz, J., Sánchez, E., Castillo, A.M. and Radulovich, R. (1994) Huertos caseros: un actividad productiva con amplia participación de la mujer. p. 151–185. In: Radulovich, R. (ed.) *Technologías Productivas para Sistemas Agrosilvopecuarios de Ladera con Sequía Estacional.* Serie Técnica No. 222, CATIE, Turrialba, Costa Rica, 185p.

Nair, N.K.R. and Kumar, B.M. (2006) Introduction. p. 1–10. In: Kumar, B.M. and Nair, N.K.R. (eds.) *Tropical Homegardens: A Time-Tested Example of Agroforestry,* Springer, NY.

Ofori, F. and Stern, W.R. (1987) Cereal legume intercopping system. *Advances in Agronomy* 41: 41–90.

Ong, C. (1994) Alley cropping. ecological pie in the sky? *Agroforestry Today* 6(3): 8–10.

Overman, A.R. and Scholtz III, R.V. (2002) *Mathematical Models of Crop Growth and Yield,* CRC Press, NY, 344p.

Palm, C.A. (1995) Contribution of agroforestry trees to nutrient requirements of intercropped plants. *Agroforestry Systems* 30: 105–124.

Paris, Q. (1992) The von Liebig hypothesis. *American Journal of Agricultural Economics* 74: 1019–1028.

Percival, J. (1922) *The Wheat Plant,* E.P. Dutton and Co., NY, 463p.

Perfecto, I., Rice, R.A., Greenberg, R. and Van der Voort, M.E. (1996) Shade coffee: a disappearing refuge for biodiversity. *BioScience* 46(8): 598–608.

Petr, J., Černý., V., Hruška, L. et al. (1988) *Yield Formation in the Main Field Crops,* Development in Crop Science (13), Elsevier, NY, 336p.

Prinsley, R.T. (1992) The role of trees in sustainable agriculture an overview. *Agroforestry Systems* 20: 87–115.

Ranganathan, R., Fafchamps, M. and Walker, T.S. (1991) Evaluating biological productivity in intercropping systems with production possibility curves. *Agricultural Systems* 36: 137–157.

Rao, M.R. (1986) Cereals in multiple cropping. p. 96–132. In: Francis, C.A. (ed.) *Multiple Cropping Systems,* MacMillian Publishing Co., NY, 383p.

Raynor, W. (1992) Economic analysis of indigenous agroforestry: a case study on Pohnpei Island, Federated State of Micronesia. p. 243–258. In: Sullivan, G.H., Huke, S.M. and Fox, J.M. (eds.) *Financial and Economic Analysis of Agroforestry Systems: Proceedings of a Workshop held in Honolulu, Hawaii, USA, July 1991.* Nitrogen Fixing Tree Association, Paia, HI, 312p.

Raza, A., Friedel, J.K. and Bodner, G. (2012) Improved water use efficiency for sustainable agriculture. p. 167–211. In: Lichfouse, E. (ed.) *Agroecology and Strategies for Climate Change,* Springer, NY, 333p.

Reinders, H.P. (2007) Multiple strategies on an organic farm in the Netherlands. *Leisa* 23(4): 32–34.

Reilly, A. (1993) *Gardening Naturally*. Better Homes and Gardens Books, Des Moines, IA, 192p.

Rham, W.L. (1853) *The Dictionary of the Farmer*, George Routledge and Company. London, 498p.

Rice, R.E., Gullison, R.E. and Reid, J.W. (1997) Can sustainable management save tropical forests? *Scientific American* 276(4): 44–49.

Roberts, I.P. (1907) *The Fertility of the Land*, The MacMillan Company. London, 421p.

Schenck, C.A. (1904) *Forest Utilization, Mensuration and Silviculture*, Biltmore?: NC, 3parts.

Schlich, W. (1910) *Schlich's Manual of Forestry*. Vol. II, 4th Edition. Bradbusy, Agnew and Co.: London, 424p.

Seran, T.H. and Brintha, I. (2010) Review of maize based intercropping. *Journal of Agronomy* 9: 136–145.

Soluri, J. (2001) Altered landscapes and transformed livelihoods: banana companies, Panama disease and rural communities on the north coast of Honduras, 1880–1950. p. 15–30. In: Flora, C. (ed.) *Interactions Between Agroecosystems and Rural Communities*, CRC Press, NY, 273p.

Steuer (2009) *ADBASE: A Multiple Objective Linear Programming Solver for Efficient Extreme Points and Unbounded Efficient Edges*, on-line: Terry.uga.edu/~rsteuer/downloads.

Strom, S. (2013a) An accidental cattle ranch. *The New York Times* CLXIII(56,318-Nov. 12): B1.

Strom, S. (2013b) Misgivings about how weed killer affects soil. *The New York Times* CLXIII(56,265-Sept. 20): B1.

Szumigalski, A.R. and van Acker, R.C. (2006) Land equivalent Ratio, light interception, and water use in annual intercrop in the presence and absence of in-crop herbicides. *Agronomy Journal* 100(4): 1154–1154.

Tilman, D., Knops, J., Wedin, D., Reich, P., Ritchie, M. and Siemann, E. (1997) The influence of functional diversity and composition on ecosystem processes. *Science* 277(29 August - 5330): 1300–1306.

Tremblay, E., Boivin, G., Brosseau, M. and Belanger, A. (2004) Toxicity effect of an insecticidal soap on the green powder aphid (*Homoptera aphididae*). *Phytoprotection* 90(1): 35–39.

Vandermeer, J. (1989) *The Ecology of Intercropping*. Cambridge University Press, Cambridge, UK, 237p.

Velasco, A., Ibrahim, M., Kass, D., Jiménez, F. and Rivas Platero, G. (1999) Concentraciones de fósforo en suelas bajo sistema silvopastoril de *Acacia mangium* con *Brachiaria humidicola*. *Agroforestería en las Americas* 6(23): 45–47.

Wiersum, K.F. (1984) Surface erosion under various tropical agroforestry systems. p. 231–339. In: O'Loughlin, C.L. and Pearce, A.J. (eds.) *Symposium on Effects of Forest land-use on Erosion and Slope Stability*, East-West Center, Hawaii, 310p.

Willey, J. (1979) Intercropping its importance and research needs Part 1. competition and yield advantage, *Field Crop Abstracts*. 32: 1–10.

Wojtkowski, P.A. (1993) Towards an understanding of tropical home gardens. *Agroforestry Systems* 24: 215–222.

Wojtkowski, P.A. (2004) *Agricultural Economics; Sustainability and Biodiversity*, Academic Press; NY, 293p.

Wojtkowski, Paul A. (2006) *Introduction to Agroecology: Principles and Practices*, Haworth Press, Binghamton, NY, 404p.

Zhang Fend (1996) Influences of shelterbelts in Chifeng, Inner Mongolia. *Unasylva* 47(185): 11–15.

Index

For Product Safety Concerns and Information please contact our EU representative GPSR@taylorandfrancis.com Taylor & Francis Verlag GmbH, Kaufingerstraße 24, 80331 München, Germany